An Illustrated Catalog of American Mushrooms

The US Department of Agriculture
Maurice B. Walters Collection

UNION
SQUARE
& CO.

NEW YORK

atelier éditions

Maurice B. Walters walking through the Alan Seeger Natural Area of Rothrock State Forest, some 390 acres that run along Standing Stone Creek in central Pennsylvania. 20 July, 1942.

A self-timed portrait Walters took of his camera setup as he prepares to photograph an instance of *Cypripedium hirsutum*—showy lady's slipper. 6 June, 1942.

Field Icons

An introduction to the USDA's
Maurice B. Walters photographs

by Maya Lydia Bushell

A man stands by the side of a road suspended in time. He is unmoving, frozen in photograph. He leans on one foot as he looks up at a sign marking the trailhead directly to his left. On his right arm he carries a large basket. We do not know its contents, or whether it yet holds anything at all. On either ide of him, the lithe trunks of hemlock trees soar toward a bright midsummer sky. Their full crowns shade the forest floor over which this man will soon be moving. This is his hunting ground. The species he pursues will not flee at the sound of his footsteps, nor cry out upon discovery. They will submit their flesh to his hand—to the gentle grasp with which he pulls them from the earth, places them in his basket, and continues in search of further specimens. At some point during the day, the camera will return to his possession and he will begin to take pictures, the click of the shutter resounding in the quiet woods as he captures one serene portrait after another. Following each new documentation, he will take out a notebook, enter a series of careful observations, and move farther into the cool shade of the trees that keep the sun's rays from falling too heavily on his path.

It is July 20, 1942, and Maurice B. Walters is making his way through the Alan Seeger Natural Area of Rothrock State Forest, some 390 acres that run along Standing Stone Creek in central Pennsylvania. He is participating in an independent foray organized by Dr. Lee Oras Overholts, a mycologist at Pennsylvania State University, whose cabin, in the borough of State College, is the host site of this excursion. It is a microcosm of the Annual Summer Foray held by the Mycological Society of America (MSA), the organization through which Walters and Dr. Overholts met, and whose purpose is, to this day, "to promote and advance the science of mycology and to foster and encourage research and education in mycology in all its aspects." It is a convivial home to "all persons with a personal or professional interest in fungi,"[2] and a group to which Walters will be a dedicated, though enigmatic, member until his death in 1979.

January 27, 1955
Mr. Maurice B. Walters
1073 Allston Road
Cleveland 21, Ohio

Dear Mr. Walters:

As you know, the Mycological Herbarium here at Beltsville has been recently designated as "The National Fungus Collections." It consists of the mycological materials of the Department of Agriculture and of the Smithsonian Institution, operating under a cooperative agreement. The Smithsonian units which we have include a general fungus collection, The Lloyd Herbarium and The W. H. Long Herbarium. We would be very happy indeed to have your mushroom plates deposited here as part of the Smithsonian Collections. I know that Mr. Swallen, Head Curator of Botany for the Smithsonian, would join me in this feeling. I saw some of your pictures at the time of the Highland's Foray and therefore have first-hand knowledge of their importance. The other institutions which you mention would be equally delighted to have them, I am sure, but I will admit a selfish desire to have them here.

Very truly yours,
    John A. Stevenson
    Principal Mycologist in Charge [3]

The United States National Agricultural Library (NAL) stands on a large swath of verdant land in Beltsville, Maryland, just outside the nation's capital. At 17 stories, it is the tallest building on the campus of the Henry A. Wallace Beltsville Agricultural Research Center. Completed in 1969, the facility, with its heavy brown-brick facade, is, while less pleasing than its Georgian Revival–style counterparts across Baltimore Avenue, neither welcoming nor decidedly uninviting. It is simply governmental.

Established on May 15, 1862, when President Abraham Lincoln signed the Department of Agriculture Organic Act, the NAL has been an integral part of the USDA since its inception. With the primary objective of disseminating and advancing scientific research on agriculture, the NAL has aided the department in that endeavor by housing and preserving for study all manner of imaginative and disparate materials—painted wax models of apples; glass vessels of natural fibers; a haunting photographic survey of the Dust Bowl; documentation of the life and work of Dr. George Washington Carver, an official Collaborator of the USDA who, during the course of his career, contributed over 1,100 specimens to the Herbarium of the US National Fungus Collections. [4]

The Collections is composed of two main facets. The first, the aforementioned Herbarium (code BPI in the global *Index Herbariorum*), is the largest collection of dried fungus specimens in the Western Hemisphere. [5] It is also the only actively curated nomenclature database focused on plant pathogenic fungi. [6] Home to over 1 million reference specimens, the Herbarium is currently under the purview of the Mycology and Nematology Genetic Diversity and Biology Laboratory. [7] Its counterpart, the Topical Files, resides with the NAL. It is here that the photographs of Maurice B. Walters have been stored, vibrant yet dormant in their archival boxes, for the past six decades.

The purpose of the Topical Files is to articulate the history of American mycology and plant pathology as it relates to the development of the US National Fungus Collections in the 19th and 20th centuries.[8] Assembled by John A. Stevenson (a noted mycologist and one of the founding members of the MSA) during his 33-year tenure as director of he Collections, included materials ranging from correspondence to mycological and phytopathological data to field and laboratory records.[9] There is also a distinct literary and visual component to the Topical Files, which hold rare manuscripts on early mycological identification and study (the oldest is the Flemish priest Franciscus van Sterbeeck's *Theatrum fungorum oft het Tooneel der Campernoelien,* published in 1675), as well as various collections of illustrations and photographs.

As the letter between Stevenson and Walters outlines, Walters first offered his photographic plates to the Collections in 1955, stating, in his response to Stevenson's acceptance of his proposed donation, "I'm glad to know that you feel that they [the plates] would be worth adding to the Smithsonian collections, and I see no reason why I should not 'earmark' them for your herbarium. [...] Of course I plan to keep them for some time yet, tho [*sic*] for just how long will depend on how things work out in the next two or three years."[10] While there is no further explanation as to why Walters held such a specific timeline for his donation, there is depth behind this comment, the essence being that perhaps his work was not yet finished— that his relationship to the photographs had not been fully explored. It may have been that he was simply not quite ready to let them go.

Four years later, writing to Stevenson on the eve of the director's retirement in December 1959, Walters once again brings up the donation of his plates. Stevenson, having written to Walters about the onslaught of collections that had been pouring in over the preceding months, elicited this response from his longtime correspondent: "Well, at least my collection of 'icons' won't add to that work, should your successor consider them worth having, and frankly I'm more than a little doubtful on the score of their worth myself."[11] Walters's verbiage is telling. His ironic self-deprecation attempts to overshadow the sincere designation of his plates as icons—representative symbols worthy of veneration, study, and display.

There are only two confirmed records of the publication of Walters's photographs in his lifetime. In 1945, mycologists Alexander H. Smith and Rolf Singer used one of Walters's images, a quartet of *Cystoderma granosum,* as the end plate of a lengthy academic article on the genus *Cystoderma.*[12] Three years prior, Dr. Fred Jay Seaver, another noted mycologist who held a number of prominent positions at the New York Botanical Garden (he was also a fellow member of the MSA), published two of Walters's plates in the supplement to his magnum opus, *The North American Cup-fungi (Operculates)* (1942).[13] Both in the *Peziza* genus, the first is a vertical portrait of two species of *Peziza pseudoclypeata* on a section of rotted wood.[14] The second is *Peziza waltersii,* which Dr. Seaver named for Walters after he discovered the hitherto unknown species near his home in Cleveland Heights, Ohio. Of the dedication, Dr. Seaver writes, "This species is named in honor of Maurice B. Walters who has collected it several years in succession on the same decaying log."[15] Upon receiving the news that Dr. Seaver wanted to name the finding after him, Walters replied, "It is very kind of you to consider naming the species for me, but I really feel that I am much too obscure a member of the clan to deserve the honor."[16]

There is a certain dissonance to the idea that a man such as Walters—who maintained relationships with some of the leading American mycologists of the 20th century, was an active member of an established society dedicated to the field, was responsible for the authorship of several new species names, and traveled to participate in forays across North America—would consider himself "too obscure" for any form of accolade. In fact, Walters's local renown was such that the Cleveland Museum of Natural History would refer those with questions about regional fungi to him.[17] (In a delightfully absurd anecdote published in an article for *Mycologia,* the MSA's journal, Walters recounts a phone call he received in October 1942 from a woman who had been directed to him by the museum. "She had been out for a walk in the woods that afternoon and had taken a few nibbles from a mushroom that she found, feeling confident that she could distinguish the edible from the poisonous ones. At the moment she was lying down and experiencing the most glorious visions of color and sounds of music, but with no feeling of discomfort whatever."[18] After advising her that she was likely in no danger, but, regardless, to call her doctor, Walters requested that she keep him updated on her condition. "She called the same evening, said the hallucinations had soon passed and that she was feeling perfectly normal again, and added that if this were the way one died of mushroom poisoning, she was all for it."[19] Later, the woman brought Walters the specimen, which he identified as *Pholiota spectabilis.*[20] It is now classified as *Gymnopilus subspectabilis* or, colloquially—big laughing mushroom.)

Just as Stevenson wrote of his "first-hand knowledge" of the importance of Walters's photographs, Dr. Seaver, too, understood the singular contribution of their contemporary's work to mycological study. Without speculating on the nature of Walters's humble self-image, it is worth mentioning that little is known about his life, and even less of his background as it relates to the natural sciences. From the few articles he had published, we know that he independently studied fungi, aquatic plants, and mosses.[21] In 1950, the bryologist Dr. William C. Steere published an article on the discovery of *Fissidens exilis* in North America—a discovery made by Walters himself, who, as with *Peziza waltersii,* had collected the species of moss in the vicinity of his home in Cleveland Heights and subsequently submitted it for identification.[22] "[He] is certainly to be congratulated," Steere writes, "for his keenness and critical ability as a collector that led to his discovery of this unusually interesting species of Fissidens, not previously reported in North America."[23] Walters was steeped in biology—he was also a regular contributor of ornithological data to the Museum of Natural History during its collection efforts in the late 1940s and early 1950s—but there is no evidence of him having ever received formal training in the discipline or its related fields.[24] He was, as Steere notes, "a well-known amateur collector," that critical third actor in the essential triangulation between herbaria and their institutional custodians.[25] Walters's last note of employment is found in the 1930 US census, where it is recorded that he was a clerk in the Purchasing Department of the Ohio Nut & Bolt Company.

Born in 1891 in Rochester, New York, the son of an English father and a French Canadian mother, the little we know of Walters's life has come from the mildly enlightening facts that public records have provided us. Through the collation of census data, marriage notices, draft registration cards, passport and visa applications, and education and housing records, Walters's portrait attempts to come into focus countless times. We know that he was a graduate of Yale University with a degree in mining engineering; a sergeant in the US Army with the 134th Field Artillery Unit, 37th Infantry Division, during the final years of World War I; and subsequently resided in Honduras and then Ecuador for a period of time while working for the South American Mining Corporation. In 1924, he married Gertrude Chamberlain of San Angelo, Texas, with whom he had two children, Gertrude and Charles. We know that Walters had blue eyes and brown hair, was of ruddy complexion, and, at 5 feet 9 inches, 117 pounds, occupied a wraithike figure befitting his slim biographical presence. [26]

Facts are evidence of existence, not existence itself. Within these records, the cycles of life, death, and employment are present. They offer an unbiased, seemingly objective account of a person's life, with clear, ruler-made lines that connect point A to point B to point C, and so on. They elucidate movements through time and space toward the markers of civil existence, and yet, when laid out before us, they fail, categorically, to follow its live edges—the curvatures of one's creativities, inner doubts, and impassioned hopes—of what we would consider life's true nature, or, rather, the

(Left) A group of Peziza waltersii, which Dr. Fred Jay Seaver named for Walters after he discovered the hitherto unknown species near his home in Cleveland Heights, Ohio. 6 June, 1942. (Right) The verso of Walters's photograph of the Peziza waltersii grouping, which includes an example of one of his color notes.

substances that drive us to maintain and externalize our existence. The
few dozen letters held at the NAL and the New York Botanical Garden aid
enormously in our understanding of Walters, for aside from the handful of
articles that he had published, there exist no other records of his voice, of
his life in his own words. For that, we must ultimately look to the language
of his photographs—the icons before which Walters prostrated himself in the
name of accuracy and artistry in a singular act of devotion that produced
an extraordinary record of American mycological study in the 20th century.

A chorus of *Armillaria amianthina* turn across the void, levitating as they
display their ochre caps and stems and tip back to reveal the fan of their gills,
as layered as a cancan dancer's skirts. Two golden fingers of *Microglossum
rufum* spring forth from a tuft of bog moss, while the basidia of a cluster
of *Calocera cornea* push up from their host of decaying wood, the tines of their
tuning fork–like structures delivering frequencies to an unknown instrument.
There is a *Marasmius glabellus* whose stem is so slender that the parachute
of its cap must be suspended in air. It seems impossible that it does not bend
under the weight of its own grace.

These are just a few of the approximately 1,000 plates that
comprise Walters's photographic archive. There are species in isolation, in
the company of pine cones, leaves, sections of bark, and loamy clumps of
soil. There are instances of riotous color that still, over half a century later,
emanate from the surface of their application—Walters, who took meticulous
color notes after photographing his specimens, hand-colored each black-
and-white print with scientific precision. In a letter to Edith K. Cash, an
associate mycologist at the Bureau of Plant Industry (predecessor of the
Animal and Plant Health Inspection Service, or APHIS), he gives a rigorous
color description of the three different shades of lavender he observed on an
unknown species in the genus *Peziza*, stating that the specimen is "externally
dull white or pale lavendar [sic], very minutely pustulate, especially toward
the margin. Hymenium with faint tinge of lavendar [sic] at first, becoming
deeper rose-lavendar [sic] to yellow-brown, slightly umbilicate."[27] On the
verso of one his prints of *Peziza waltersii*, the species Dr. Seaver named after
Walters, he reveals his dissatisfaction with his coloring job, leaving himself
the parenthetical note, "The yellow should be a bit duller." As Shannon
Dominick, a biologist with the Mycology and Nematology Genetic Diversity
and Biology Laboratory, reflected in response to Walters's photographs,
"The art *is* the science."

All of these plates represent the rich mycological universe into which
Walters repeatedly entered. The majority of the species he documented were
found on the North Chagrin Reservation, a 1,600-acre section of the Cleveland
Metropark system that includes forests, meadows, and wetlands. This diverse
terrain, notable, according to Walters, for its "unusual richness," was perennial
in its offerings to those patient enough to return, year after year, and chronicle
its abundances.[28]

Walters supplemented his regional study with MSA Forays to Maine,
Vermont, North Carolina, and Québec Province, among other locales. There
are remarkable photographic records of these trips, both in mycological
documentation and the more narrative studies that emerged from the genial

atmosphere of these gatherings. Men and women, boys and dogs, are posed with the most fantastic finds—enormous specimens that fill laps, hands, and arms and dwarf their canine companions. There are shots of fungi in situ, their proportions given scale by the items available in someone's pocket—a wooden pipe and a pocket watch rest on the substantial cap of a *Boletus felleus,* 10½ inches in diameter. (There are a few photographs, such as this, whose Surrealist dispositions are so strong that we half expect the long ends of Salvador Dalí's mustache to come poking into frame.) Amid these narrative illustrations, we also find the sole visual reference to Walters's camera setup, an integral yet elusive facet of his mycological work of which there is little to no written description. In this plate, Walters kneels beside his camera, the device shrouded in dark cloth as he prepares to photograph an occurrence not of fungi but, rather, an elegant *Cypripedium hirsutum*—showy lady's slipper.

There is a resonance to this last image—that of a man whose work is the primary lens through which we understand his modes of study, his motivations and pleasures, the ways in which he exercised so much of his existence, but about whom we admit to knowing relatively little. Dark lacunae threaten to black out our picture of Walters, and yet here, on this miraculous day, June 6, 1942, he is entirely whole.

A man stands by the side of a road suspended in time. He is unmoving, frozen in photograph. He leans on one foot as he looks up at a sign marking the trailhead directly to his left. On his right arm he carries a large basket. We do not know its contents, or whether it yet holds anything at all. On either side of him, the lithe trunks of hemlock trees soar toward a bright midsummer sky. Their full crowns shade the forest floor over which this man will soon be moving. This is his hunting ground. The species he pursues will not flee at he sound of his footsteps, nor cry out upon discovery. They will submit their flesh to his hand—to the gentle grasp with which he pulls them from the earth, places them in his basket, and continues in search of further specimens.

Another group of *Peziza waltersii.* Walters's photograph
was published in Dr. Seaver's supplement to his magnum opus,
The North American Cup-fungi (Operculates) (1942).

(Top) One of Walters's humorous tableaus: a *Boletus felleus* (more commonly known as *Tylopilus felleus*) giving shelter to a family of skunks and a gremlin. 15 July, 1950. *(Bottom)* Walters sitting outside the cabin of mycologist Dr. Lee Oras Overholts with a table full of mushroom specimens spread out before him. State College, Pennsylvania. 21 July, 1942.

14

(Top) Walters leading a "Mushroom Walk" for members of the Cleveland Natural Sciences Club at South Chagrin Metropark. Willoughby Hills, Ohio. Circa 21 September, 1952. *(Bottom)* Walters with fellow mycology enthusiasts Ruby and Charley Margachs during a hike in North Chagrin Metropark. Gates Mills, Ohio. 14 June, 1943. *(Page overleaf)* A trail leading through the Alan Seeger Natural Area near State College, Pennsylvania. 20 July, 1942.

Walters's Hunting Grounds

As an active member of the Mycological Society of America (MSA), Walters was a consistent participant in the organization's Annual Summer Forays, often traveling with his family to a variety of locales across the United States and Canada. While the majority of the species Walters photographed were found and then documented in Ohio, nearby to his home in Cleveland Heights (the North Chagrin Reservation was a particularly fruitful hunting ground), the MSA Forays provided unique opportunities to engage with diverse mycological landscapes farther afield in a convivial atmosphere that leant itself to study, leisure, and play—for both adults and children. The locations of these forays, both officially and independently organized by the MSA and Walters, respectively, are indexed here for geographical context.

FERN LAKE BOG, BURTON TOWNSHIP, OHIO

Walters frequented this 14.6-acre area of the Burton County Wetlands, now under the stewardship of the Cleveland Museum of Natural History's Natural Areas Program, the goal of which is to create a system of nature preserves that best represents the broad range of habitats in northern Ohio.

HATCH-OTIS WILDLIFE SANCTUARY, WILLOUGHBY HILLS, OHIO

An 81-acre sanctuary established between 1944 and 1947 on land donated by the Burroughs Nature Club to the Cleveland Bird Club, of which Walters was an active member. Recognized as an Old-Growth Forest by the Old-Growth Forest Network, the woodlands include American beech, northern red oak, sassafras, sugar maple, and hemlock trees.

HOLDEN ARBORETUM, KIRTLAND, OHIO

Established in 1931, the Holden Arboretum, now integrated with the Cleveland Botanical Garden and known as the Holden Forests & Gardens, was originally under the purview of the Cleveland Museum of Natural History until its separation from the institution in 1952. Located just east of Cleveland, within Lake and Geauga Counties, the arboretum encompasses trails, gardens, lakes, and meadows.

NORTH CHAGRIN RESERVATION, GATES MILLS, OHIO

Established in the 1920s, the North Chagrin Reservation is a combination of both outdoor recreation areas and wildlife sanctuaries. With large areas of both woodlands and wetlands, the reservation offers distinct opportunities to observe the region's native flora and fauna, such as wild-flowers and waterfowl. Walters often frequented the Strawberry and Forest Lane areas, where he found and documented many different fungal species.

SHAKER LAKE, THE SHAKER PARKLANDS, CLEVELAND HEIGHTS, OHIO

One of two lakes created in the mid-19th century by the North Union Shaker Community, which dammed the Doan Brook Watershed to generate waterpower for nearby wool, saw-, and gristmills. In 1895, this 279-acre region was donated to the City of Cleveland, which then leased the area (including Shaker Lake, rebuilt by the landscape architect Ernest W. Bowditch) to the cities of Shaker Heights and Cleveland Heights, where Walters and his family resided.

SOUTH CHAGRIN RESERVATION, WILLOUGHBY HILLS, OHIO

A heavily wooded area with hemlock forests, sand-stone ledges, ravines, and bodies of water, including cold-water creeks and the Chagrin River, a designated Scenic River.

PERSONAL TRIPS

ALAN SEEGER NATURAL AREA, HUNTINGDON, PENNSYLVANIA

Located in the Rothrock State Forest in central Pennsylvania, this 390-acre Natural Area runs along Standing Stone Creek and includes both virgin white pine and hemlock woodlands, with some examples of the latter species measuring over four feet DBH (diameter at breast height, a standard measurement of a tree's diameter). Other tree species, including those in the remnants of the Area's old-growth forest, consist of white oak, red maple, white pine, pignut hickory, black gum, and black birch.

WHITE MOUNTAINS, NEW HAMPSHIRE

An extensive mountain range that covers approximately one quarter of the state of New Hampshire, as well as a small section of western Maine. It includes Mount Washington, which is the highest peak in the northeastern United States, standing at 6,288 feet (1,917 meters). A portion of the Appalachian Trail crosses the 49-peak range from southwest to northeast.

STATION TOURISTIQUE DUCHESNAY, QUÉBEC, CANADA (23–28 AUGUST, 1938)

Located outside of Québec City in the La Jacques–Cartier Regional County Municipality, Station Touristique Duchesnay is a nature center that sits on the northern, western, and eastern shores of Saint-Joseph Lake. In the early 20th century, the station was home to a Forest Rangers School, which conducted important field research related to forest pathology.

GREAT SMOKY MOUNTAINS NATIONAL PARK, TENNESSEE (16–20 AUGUST, 1939)

Currently the most visited National Park in the US, Great Smoky Mountains, which sits on the border between North Carolina and Tennessee, is renowned for its diversity of plant and animal life. The park's highest peak is Kuwohi (formerly Clingmans Dome), with an elevation of 6,643 feet (2,025 meters). Its Cherokee name translates to "mulberry place."

MOUNT KATAHDIN, MAINE (21–24 AUGUST, 1940)

Located in Baxter State Park, Mount Katahdin (which translates to "Great Mountain" in Penobscot) is Maine's highest mountain. With an elevation of 5,269 feet (1,606 meters), Baxter Peak is the northern terminus of the Appalachian National Scenic Trail. The mountain is held sacred by the Maliseet, Micmac, Passamaquoddy, and Penobscot nations.

MACDONALD COLLEGE, SAINTE-ANNE-DE-BELLEVUE, QUÉBEC, CANADA (26–29 AUGUST, 1941)

First opened in 1907, Macdonald College is home to McGill University's Faculty of Agricultural and Environmental Sciences (FAES). It is located in the West Island region of the Island of Montréal.

UNIVERSITY OF WEST VIRGINIA, MORGANTOWN, WEST VIRGINIA (30 AUGUST–2 SEPTEMBER, 1946)

WVU is a public land-grant research university. Founded as the Agricultural College of West Virginia in 1867, the university has had strong agricultural and forestry programs since its inception.

HIGHLANDS BIOLOGICAL RESEARCH STATION, HIGHLANDS, NORTH CAROLINA (2–7 SEPTEMBER, 1947)

The Highlands Biological Station (HBS) is multi-campus center of Western Carolina University, the mission of which is to foster research and education focused on the rich natural heritage of the Highlands Plateau, known locally as the "Land of Waterfalls." Due to the region's high annual rainfall, the area has developed distinct and verdant microclimates, which appeal to botanists and their research interests.

UNIVERSITY OF VERMONT, BURLINGTON, VERMONT (JUNE 1951)

Officially the University of Vermont and State Agricultural College, UVM is a public land-grant research university known for both its liberal arts and science programs. The university was founded in 1791, the same year Vermont became the 14th US state, and its Latin name, Universitas Viridis Montis, translates to "University of the Green Mountains."

MSA members in the midst of collecting specimens at Macdonald College in Sainte-Anne-de-Bellevue, Québec, Canada, during the Annual Summer Foray. Left to right: Mosely (one of the Foray hosts), Dr. Groves, Dr. Dearness, Mrs. Groves, and Dr. Conner's daughter. 1941.

(*Top*) The Walters family—Maurice; his wife, Gertrude; and their two children, Gertrude and Charles—stopped for lunch at Indian Gap. They are returning from what was known at the time as Clingmans Dome, but which, in 2022, was renamed Kuwohi, meaning "mulberry place," the original name given to the area by the Cherokee. This photo was taken during the MSA Summer Foray at Great Smoky Mountains National Park (16–20 August, 1939). (*Bottom*) Walters standing at the edge of Lake Ambajejus, which he described as a "moosey little lake," close to the cabin where he and his family were staying during the Mount Katahdin Foray. 24 August, 1940.

(*Top*) Walters documents his stay at Dr. Overholts's cabin during their independently organized foray, annotating this image as an "appropriate view of a mycologists's cabin." Alan Seeger Monument, Pennsylvania. 21 July, 1942. (*Bottom*) A merry band of MSA members in the midst of collecting specimens during the Summer Foray to Macdonald College in Sainte-Anne-de-Bellevue, Québec, Canada. 26–29 August, 1941. Walters's daughter, Gertrude, is pictured second from the right.

(Top) The Walters family pictured outside their cabin during the Summer Foray to Mount Katahdin, Maine. 20–24 August, 1940. *(Bottom)* The view across Lake Ambajejus from the cabin where the Walters family stayed during the Summer Foray to Mount Katahdin. Walters's children, Gertrude and Charles, are pictured in the rowboat. 21 August, 1940.

Specimens

BOLETUS
No 5465
Boletus edulis
(Penny bun, cep, porcino)
North Chagrin Reservation
Gates Mills, Ohio
Date unknown

The *Boletus edulis* grows in coniferous forests, forming symbiotic associations with the surrounding trees by enveloping their roots in fungal tissue.

Known commonly as the penny bun, cep, or porcino mushroom, with further common names varying by region, it is edible and very popular in cuisines (its Latin epiphet *edulis* means "edible"), most often prepared in risottos, pastas, and soups. It can also be dried and, unusually, pickled.

BOLETUS
No 5475
Boletus fumosipes Pk
North Chagrin Reservation
Gates Mills, Ohio
19 July, 1941

CLITOCYBE
No 5572
Clitocybe multiformis Pk
North Chagrin Reservation
Willoughby Hills, Gates Mills, Mayfield, Ohio
8 September, 1942

CLITOCYBE
No 5571
Clitocybe multiceps Pk
North Chagrin Reservation
Gates Mills, Ohio
2 November, 1941

Walters notes having found the specimen "in grass in tree lawn along Cedar Road."

29

CORTINARIUS
No 5638
Cortinarius lilacinus Pk
North Chagrin Reservation
Willoughby Hills, Gates Mills, Mayfield, Ohio
1 October, 1934

Walters notes "North Chagrin Found in rather wet ground in open woods used by cattle. Gills crowded close, uneven adnate, notched, ventricose Sores buff." He also notes that this is one of his earliest photographs.

AMANITA
No 5393
Amanita cinereoconia Atk
North Chagrin Reservation
Willoughby Hills, Gates Mills, Mayfield, Ohio
14 August, 1935

AMANITA
No 5383
Amanita brunnescens Atk
(Brown American star-footed amanita,
cleft-footed amanita)
North Chagrin Reservation, Gates Mills, Ohio
9 July, 1939

The *Amanita brunnescens* is sometimes known as the cleft-footed amanita
as a result of its recognizable cleft-foot base. Its edibility is unknown; it
may be poisonous, and was for many years presumed to be the death cap
mushroom due to a similarity in appearance to the *Amanita phalloides*.

LEPIOTA
No 5879
Lepiota naucina Fr
North Chagrin Reservation
Willoughby Hills, Gates Mills, Mayfield, Ohio
30 September, 1938

BOLETUS
No 5456
Boletus alboater Schw
Holden Arboretum
Kirtland, Ohio
14 August, 1937

34

ENTOLOMA
No 5693
Entoloma cyaneum Pk
Thompson Ledges Township Park
Thompson, Ohio
29 September, 1938

POLYPORUS
No 6429
Polyporus radicatus (Schw)
Holden Arboretum
Kirtland, Ohio
8 August, 1937

FLAMMULA
No 5712
Flammula lenta Fr
North Chagrin Reservation
Willoughby Hills, Gates Mills, Mayfield, Ohio
7 October, 1938

COLLYBIA
No 5594
Collybia maculata A & S
McGill University
Sainte-Anne-de-Bellevue, Québec, Canada
25 August, 1941

38

TRICHOLOMA
No 6214
Tricholoma sulphureum Fr
North Chagrin Reservation
Willoughby Hills, Gates Mills, Mayfield, Ohio
27 September, 1945

(Top) ARMILLARIA, No 5439. *Armillaria granosa (Morg) Kauff.* North Chagrin Reservation, Willoughby Hills, Gates Mills, Mayfield, Ohio. 5 September, 1941. *(Bottom)* BOLETUS, No 5457. *Boletus auriporus Pk.* North Chagrin Reservation, Gates Mills, Ohio. 24 July, 1943.

BOLETUS
No 6342
Boletus miniato-olivaccus Fr
North Chagrin Reservation
Willoughby Hills, Gates Mills, Mayfield, Ohio
26 July, 1950

41

CANTHARELLUS
No 5520
Cantharellus cibarius Fr
（Golden chanterelle）
North Chagrin Reservation
Gates Mills, Ohio
12 July, 1943

Walters noted that the cap color was off in this study.

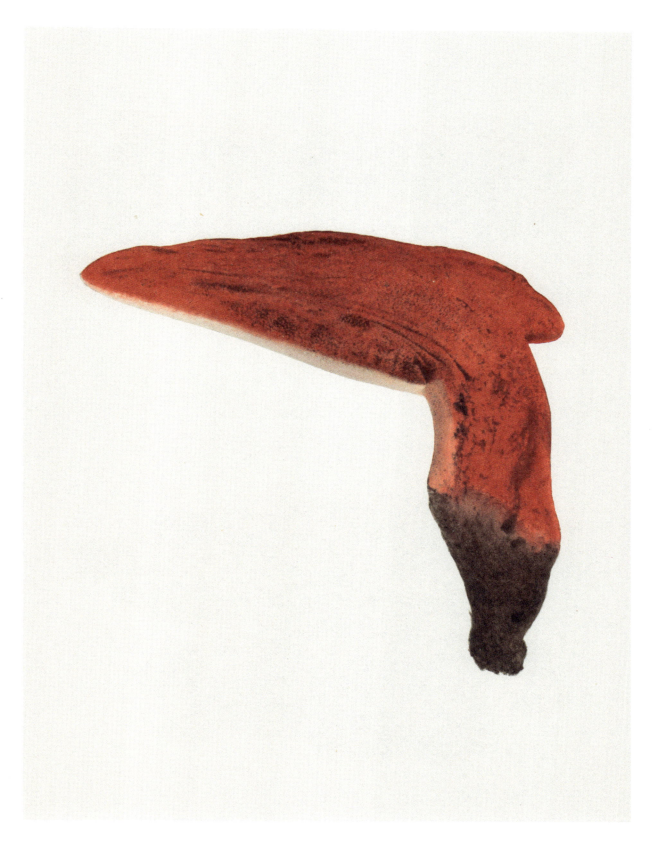

MISC
No 5709
Fistulina hepatica (Huds) Fr
Holden Arboretum
Kirtland, Ohio
14 August, 1935

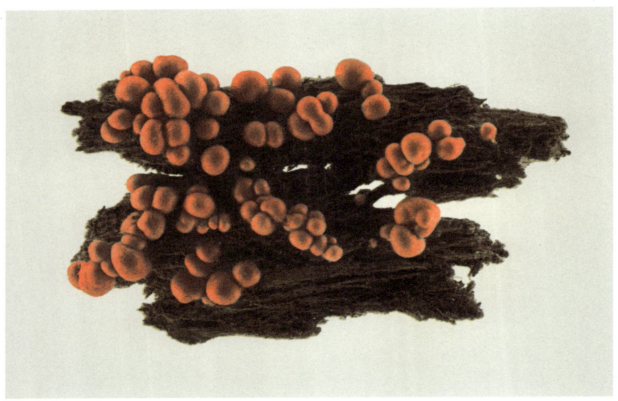

(Top) ARCYRIA, No 5443. *Arcyria denudata (L) Sheldon.* North Chagrin Reservation, Willoughby Hills, Gates Mills, Mayfield, Ohio. 10 September, 1942. *(Bottom)* LYCOGALA, No 5889. *Lycogala epidendrum (Buxb) Fr* (wolf's milk). North Chagrin Reservation, Willoughby Hills, Gates Mills, Mayfield, Ohio. 28 May, 1939.

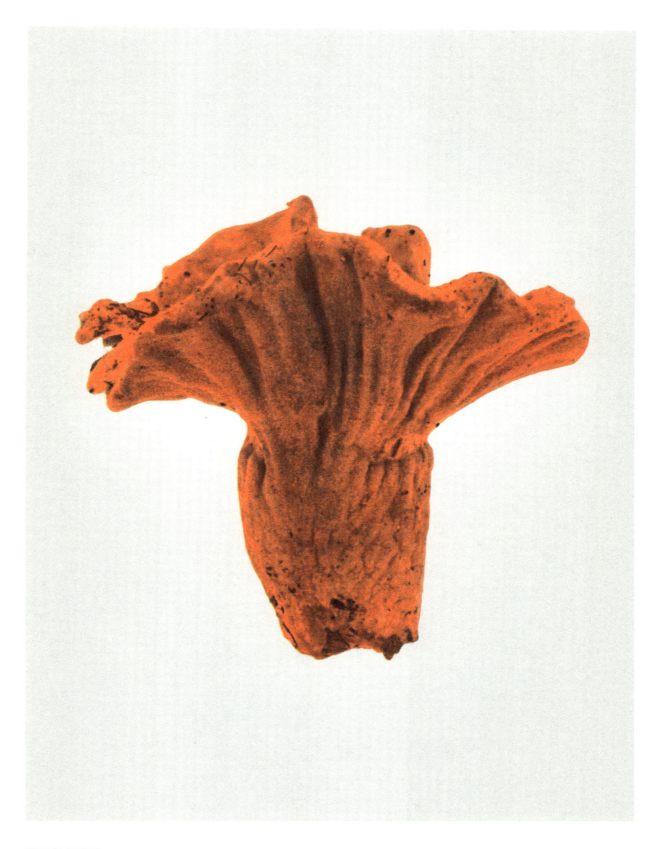

HYPOMYCES
No 5805
Hypomyces lactifluorum (Schw)
(Lobster mushroom)
North Chagrin Reservation
Gates Mills, Ohio
9 August, 1943

Hypomyces lactifluorum, commonly known as the lobster mushroom, is a parasitic fungus that grows on other mushrooms, turning them a shade of reddish-orange resembling a cooked lobster's shell. Lobster mushrooms are edible and widely foraged, with a firm texture and shellfish flavor.

CANTHARELLUS .
No 5521
Cantharellus floccosus Schw
(Scaly vase, shaggy/scaly/wooly chanterelle)
Holden Arboretum
Kirtland, Ohio
11 July, 1938

The *Cantharellus floccosus* forms symbiotic relationships with conifers, and occurs in coniferous forests in North America in late summer and fall. Commonly known as the scaly vase or the wooly chanterelle, the mushroom can be poisonous to some who eat it, causing nausea, vomiting, and diarrhea, though there is evidence of others consuming the fungi without incident. The mushroom is commonly eaten by the Sherpa people in Nepal, as well as across Mexico.

CANTHARELLUS
No 5522
Cantharellus minor Pk
Quarry Park North
South Euclid, Ohio
10 July, 1943

The *Cantharellus minor* is a very small member of the *Cantharellus* genus, with its cap ranging in size from 0.5 to 3 centimeters. The mushroom is native to North America and grows among oaks and mosses. It is edible but not commonly eaten due to its insubstantial size.

CANTHARELLUS
No 5519
Cantharellus cibarius Fr
(Golden chanterelle)
Quarry Park North
South Euclid, Ohio
10 July, 1943

Walters noted that this specimen was "Probably only a slightly more scaly, and somewhat darker-centered, form of the contrelle [*sic*]."

The *Cantharellus cibarius* or golden chanterelle is a commonly consumed mushroom, recognized by its yellow color and faint apricot odor and flavor. However, caution is advised when foraging this mushroom due to its similarities to the *Omphalotus illudens* (eastern jack-o'-lantern) and the *Cantharellus aurantiacus* (false chanterelle), both of which are poisonous.

CLITOCYBE
No 5551
Clitocybe catina Fr
North Chagrin Reservation
Gates Mills, Ohio
9 August, 1942

MARASMIUS
No 5911
Marasmius scorodonius Fr
North Chagrin Reservation
Willoughby Hills, Gates Mills, Mayfield, Ohio
28 June, 1951

Marasmius scorodonius is found on conifer needles, sticks, and bark across Europe, North America, North Africa, and Asia, from summer to late fall. Its taste is described as unpleasant, but it can be used to add a garlic flavor in cooking, and has a strong garlic odor.

COPRINUS
No 5623
Coprinus plicatilis Fr
(Pleated inkcap)
Holden Arboretum
Kirtland, Ohio
8 October, 1938

Known commonly as the pleated inkcap, *Coprinus plicatilis* is a saprotrophic mushroom found across North America and Europe. The fungi's bodies grow at night following a rain, and self-decompose after dispersing their spores or otherwise eventually collapse under the weight of their own caps.

LEPIOTA
No 5864
Lepiota acerina Pk
North Chagrin Reservation
Willoughby Hills, Gates Mills, Mayfield, Ohio
29 August, 1937

52

MARASMIUS
No 5900
Marasmius erythropus Fr
North Chagrin Reservation
Willoughby Hills, Gates Mills, Mayfield, Ohio
14 July, 1937

CLITOCYBE
No 5552
Clitocybe clavipes (Pers) Quel
(Club foot)
North Chagrin Reservation
Willoughby Hills, Gates Mills, Mayfield, Ohio
27 September, 1939

The specific epiphet of this mushroom emerged from the Latin *clava*, meaning "club," and *pes*, meaning "foot," and is commonly known as the club foot or club-footed clitocybe. *Clitocybe clavipes* is found in deciduous and conifer forests. While some guides have categorized the fungi as edible, it has been compared to eating wet cotton, and contains toxins that can cause symptoms including flushing and swelling when consumed alongside alcohol.

CLITOCYBE
No 5561
Clitocybe infundibuliformis Fr
North Chagrin Reservation
Willoughby Hills, Gates Mills, Mayfield, Ohio
3 September, 1940

(Top) ARMILLARIA, No 5432. *Armillaria cinnabarina (Fr) Kauff.* North Chagrin Reservation, Gates Mills, Ohio. 17 September, 1942. *(Bottom)* HYDNUM, No 5763. *Hydnum pulcherrimum B & C.* North Chagrin Reservation, Willoughby Hills, Gates Mills, Mayfield, Ohio. 17 August, 1947.

CLAUDOPUS
No 5527
Claudopus nidulans Fr
(Mock oyster, orange oyster)
North Chagrin Reservation
Holden Arboretum, Kirtland, Ohio
5 September, 1940

The *Claudopus nidulans* has a rounded cap, which can be up to 10 centimeters wide, and an unpleasant odor, which has been likened to rotten eggs or cabbage. While the mushroom is edible, this smell typically puts people off. The common name of mock oyster references *C. nidulans*'s visual simularity to *Pleurotus ostreatus*, the oyster mushroom.

CLITOCYBE
No 5547
Clitocybe catina Fr
North Chagrin Reservation
Willoughby Hills, Gates Mills, Mayfield, Ohio
3 September, 1940

HYGROPHORUS
No 5782
Hygrophorus miniatus Fr var. subluteus
North Chagrin Reservation
Willoughby Hills, Gates Mills, Mayfield, Ohio
9 July, 1938

Walters seemed unsure of the identification of this specimen, adding a later note that it was doubtful that this was indeed "subluteus."

PLEUROTUS
No 6368
Pleurotus ostreatus (Jacq)
(Oyster mushroom)
North Chagrin Reservation
Willoughby Hills, Gates Mills, Mayfield, Ohio
24 June, 1935

The oyster mushroom (*Pleurotus ostreatus*) is a common edible species of saprobe—fungi that live off of decaying organic matter. They are found growing in tiers or clusters on the sides of dead or dying deciduous trees (especially beech), with each individual fruiting body appearing as a gray-blue, oyster-shell-like form—hence their colloquial name. Humorously, the general epithet *Pluerotus* translates from the Latin as "side-ear," which refers to the sideways direction of the cap's growth in relation to its stem. *P. ostreatus* can be found year round, though the fungus often appears after a cold snap, as the weather event triggers its fruiting. Oyster mushrooms are capable of breaking down complex materials such as cellulose and lignin, which, together, form the chief constituent of wood. As the fungus works to break these materials down, vital elements and minerals are released back into the ecosystem in a usable form for other organisms and plants. *P. ostreatus* is also one of approximately 700 known nematophagous mushrooms—the mycelia of these species are able to kill and digest nematodes (roundworms), a behavior that classifies the fungus as carnivorous. Oyster mushrooms have long been a part of various European and Asian cuisines, with cultivation of *P. ostreatus* and other related species popular in Hungary, China, and Japan. Taiwan is known for its cultivation of *P. cystidiosus*—the abalone mushroom.

PLEUROTUS
No 6075
Pleurotus sulfuroides Pk
North Chagrin Reservation
Willoughby Hills, Gates Mills, Mayfield, Ohio
18 September, 1943

PLEUROTUS
No 6073
Pleurotus serotinus Fr
North Chagrin Reservation
Willoughby Hills, Gates Mills, Mayfield, Ohio
25 October, 1937

PLEUROTUS
No 6074
Pleurotus sulfuroides Pk
North Chagrin Reservation
Willoughby Hills, Gates Mills, Mayfield, Ohio
18 September, 1940

PLEUROTUS
No 6076
Pleurotus ulmarius Fr
North Chagrin Reservation
Willoughby Hills, Gates Mills, Mayfield, Ohio
2 October, 1937

GALERA
No 5727
Galera (conocybe) pygmecaffinis Fr
North Chagrin Reservation
Willoughby Hills, Gates Mills, Mayfield, Ohio
16 November, 1941

COPRINUS
No 6350
Coprinus sylvaticus Pk
North Chagrin Reservation
Willoughby Hills, Gates Mills, Mayfield, Ohio
16 September, 1949

COPRINUS
No 5621
Coprinus micaceus Fr
(Mica cap, glistening inky cap, shiny cap)
North Chagrin Reservation
Gates Mills, Ohio
29 June, 1942

The specific epithet *micaceus* comes from the Latin word *mica*, meaning "grain of salt." The common name of "shiny cap" comes from the glistening particles on the fungi's cap. These saprotrophic fungi derive nutrients from dead and decaying matter. It is an edible mushroom, commonly used in omelets and sauces, though it has a delicate flavor that can be easily overpowered in cooking.

COPRINUS
No 5622
Coprinus micaceus
(Mica cap, glistening inky cap, shiny cap)
Location unknown
Date unknown

MISC
No 6189
Strobilomyces strobilaceus (Scop) Berk
North Chagrin Reservation
Willoughby Hills, Gates Mills, Mayfield, Ohio
23 July, 1935

Walters nicknamed this specimen the "Pine Cone Boletus" in his notes.

Old man of the woods (*Strobilomyces strobilaceus*) is a woodland species of mushroom in the Bolataceae family that, while most commonly found in deciduous forests, has also been reported under conifers. This bolete (a type of mushroom with a cap that often features spongy pores instead of gills on its underside) is notable for the wool scales that cover both its cap and stem, thus accounting for the etymology of the species: The generic name (genus) *Strobilomyces* comes from the ancient Greek word *strobilos*, meaning pine cone, from which *strobilaceus*, the specific epithet (species), also derives. Old man of the woods was first described scientifically by Italian naturalist Giovanni Antonio Scopoli in 1770, when he classified the mushroom as a species of *Boletus*—a genus originally broadly defined in 1753 by Carl Linneaus (the Swedish biologist and physician who formalized binomial nomenclature) as containing all fungi with hymenial pores instead of gills.

BOLETUS
No 5470
Boletus felleus (Bull) Fr
（Bitter bolete）
North Chagrin Reservation
Willoughby Hills, Gates Mills, Mayfield, Ohio
31 July, 1941

Walters noted that this speciment was "North Chagrin Met Park – The
type that grows on decayed hemlock logs. The only Boletus growing on
hemlock I have seen mentioned is B. mirabilis (Smith – Mushroom Hunter's
Field Guide, p. 82), but description in Myc. 4, p. 98 of Murrill's sp. nov. is
very different."

72

BOLETUS
No 5480
Boletus luteus
(Slippery jack, sticky bun)
North Chagrin Reservation
Willoughby Hills, Gates Mills, Mayfield, Ohio
Date unknown

The *Boletus luteus* mushroom, commonly known as slippery jack, is, like most boletus mushrooms, edible, though considered less high-quality than others of the genus such as the *Boletus edulis*. It is considered a delicacy in Slavic cultures, and is typically fried and used in stews or soups.

(Top left) BOLETUS, No 5471. *Boletus felleus (Bull) Fr* (bitter bolete). North Chagrin Reservation, Gates Mills, Ohio. 18 July, 1945. *(Top right)* BOLETUS, No 5494. *Boletus separans Pk*. North Chagrin Reservation, Gates Mills, Ohio. 14 August, 1942. *(Bottom left)* BOLETUS, No 6335. *Boletus edulis (Bull) Fr.* North Chagrin Reservation, Willoughby Hills, Gates Mills, Mayfield, Ohio. 23 July, 1947. *(Bottom right)* BOLETUS, No 6336. *Boletus eximius Pk* (lilac brown bolete). Highlands, North Carolina. 6 September, 1947.

BOLETUS
No 5469
Boletus felleus (Bull) Fr
(Bitter bolete)
North Chagrin Reservation
Willoughby Hills, Gates Mills, Mayfield, Ohio
15 June, 1941

The *Boletus felleus* grows in deciduous and coniferous woodland, typically fruiting under beech or oak trees. *Fell* is Latin for "bile," referring to the mushroom's bitter, bile-like taste, which has resulted in the fungi being known by the common name bitter bolete. Its flavor worsens when cooked, and can spoil the taste of an entire meal. In the book *Edible Wild Mushrooms of North America* (1992), the author writes of *Boletus felleus:* "Even when cooking, it smells terrific, but one taste of the Bitter Bolete would not only disappoint but perhaps depress the novice mushroom hunter."

MISC
No 6004
Paxillus atrotomentosus (Batsch)
North Chagrin Reservation
Willoughby Hills, Gates Mills, Mayfield, Ohio
14 July, 1935

Walters nicknamed this specimen the "Blackfooted Paxillus."

The velvet-footed tap (*Tapinella atrotomentosa*) or velvet roll-rim is, though gilled, a member of the pored mushroom order Bolatales. First described by the German naturalist August Batsch in his book *Elenchus Fungorum* (Discussion of Fungi) (1783), its current name was given to it by the Czech mycologist Josef Šutara in 1992. With its trademark rolled-rim cap, the mushroom's stem is covered with velvety dark brown or black fur, hence the etymology of the species—the specific epithet *atrotomentosa* derives from the Latin word *atrotomentosus*, meaning "black-haired." Velvet-footed taps contain several compounds, such as paxillosterone, that act as deterrents to feeding by insects.

(*Top*) BULGARIA, No 5503. *Bulgaria rufa Schw* (rubber cup, hairy rubber cup). North Chagrin Reservation, Willoughby Hills, Gates Mills, Mayfield, Ohio. 10 June, 1942. (*Bottom*) CORYNE, No 5645. *Coryne sarcoides (Jacq) Tul*. North Chagrin Reservation, Willoughby Hills, Gates Mills, Mayfield, Ohio. 10 October, 1938.

DISCINA
No 5684
Discina perlata
(Pig's ears)
Location unknown
Date unknown

Discina perlata has a distinctive wrinkled ear shape, hence its common name of pig's ears. It grows in coniferous areas both alone and in groups across temperate North America.

(Top) AURICULARIA, No 5444. *Auricularia auricula – Judae (L) Berk* (wood ear, jelly ear). North Chagrin Reservation, Gates Mills, Ohio. 1942. *(Bottom)* ALEURINA, No 5380. *Aleurina atrovinosa (Cooke) Seaver.* North Chagrin Reservation, Willoughby Hills, Gates Mills, Mayfield, Ohio. 12 July, 1942.

DISCINA
No 5682
Discina leucoxantha
Location unknown
Date unknown

(Top) CLITOCYBE, No 5549. *Clitocybe cartilaginea Burl – Bres.* North Chagrin Reservation, Gates Mills, Ohio. 18 September, 1940. *(Bottom)* CLITOCYBE, No. 5570. *Clitocybe multiceps Pk.* North Chagrin Reservation, Gates Mills, Ohio. 18 September, 1940.

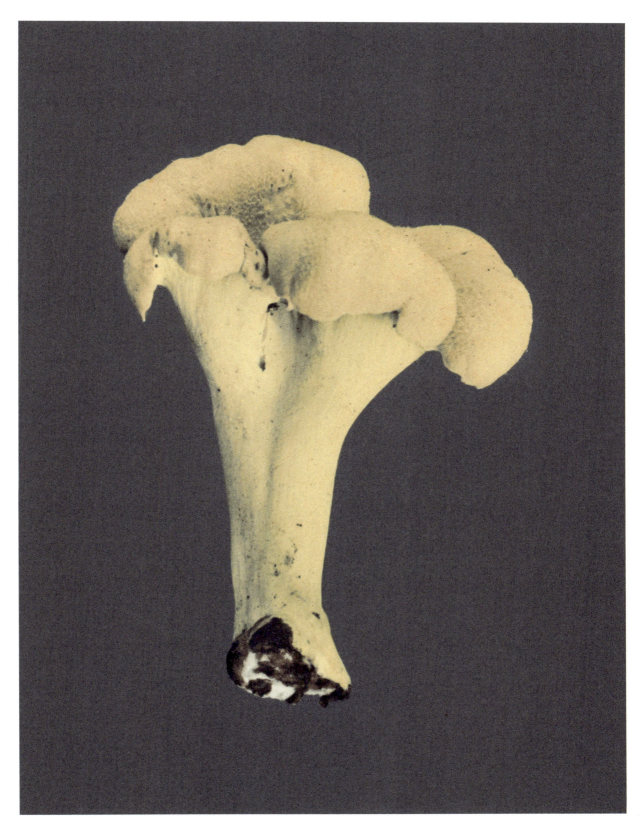

CRATERELLUS
No 5647
Craterellus cantharellus (Schw) Fr
North Chagrin Reservation
Willoughby Hills, Gates Mills, Mayfield, Ohio
1 August, 1938

CANTHARELLUS
No 5516
Cantharellus aurantiacus Fr
(False chanterelle)
North Chagrin Reservation
Willoughby Hills, Gates Mills, Mayfield, Ohio
3 September, 1944

Cantharellus aurantiacus have a golden orange cap with orange gills, and flesh ranging in color from white to yellow to orange. The mushroom's common name, false chanterelle, references its visual similarities to *Cantharellus cibarius*, the golden chanterelle. This fungi is distinguished by its velvety cap, deeper orange color, and absence of the golden chanterelle's recognizable apricot smell. It is considered poisonous, though it has traditionally been eaten by the Zapotec and Tepehuán peoples of Mexico, the latter typically boiling the mushroom and eating it with cheese.

CLITOCYBE
No 5558
Clitocybe illudens
(Eastern jack-o'-lantern mushroom)
North Chagrin Reservation
Willoughby Hills, Gates Mills, Mayfield, Ohio
Date unknown

The *Clitocybe illudens* tends to be found on the base of decaying tree stumps in North America. It is sometimes confused with edible forms of chanterelles, but can be distinguished by its fleshier appearance and tendency to form in large clusters. This distinction is important, as the eastern jack-o'-lantern is a highly poisonous mushroom, causing vomiting and cramps when consumed by humans.

CLAVARIA
No 5531
Clavaria cineroides Atk
North Chagrin Reservation
Willoughby Hills, Gates Mills, Mayfield, Ohio
16 August, 1942

CLAVARIA
No 5530
Clavaria aurea (Schaeff) var. australis
Holden Arboretum
Kirtland, Ohio
7 July, 1937

Walters noted that "Color should have been about 10-I-7, Spores slightly paler in mass, 9 × 4, granular, slightly and minutely rough."

A coral mushroom in the Gomphaceae family, the *Clavaria aurea* is an edible mushroom found in North America and Europe.

CLAVARIA
No 5538
Clavaria kunzei Fr
(White coral)
Holden Arboretum
Kirtland, Ohio
14 August, 1937

The common name of *Clavaria kunzei,* white coral, refers to the mushroom's resemblance to marine coral with its white, branched structure. Though edible, this fungus has been described as odorless and flavorless.

88

No 6286
Mycological Society Summer Foray, Great Smokey
Mountains N.P.
Great Smoky Mountains National Park
Gatlinburg, Tennessee
17 August, 1939

CLAVARIA
No 5536
Clavaria fusiformis (Sowerby)
Holden Arboretum
Kirtland, Ohio
12 August, 1937

CLAVARIA
No 5537
Clavaria fusiformis (Sowerby)
(Golden spindles, golden fairy spindles, spindle-
shaped yellow coral)
North Chagrin Reservation, Gates Mills, Ohio
9 July, 1939

Walters noted: "Cedar – Gates Mills woods – Spores spherical, 6–7. Oil drop nearly filling spore. Very brittle. No taste nor odor. Spores white in mass. Dried out paler but with tips darker."

The fruit bodies of *Clavaria fusiformis* form bright yellow, cylindrical clusters, typically in agriculturally untouched grassland and amid woodland litter.

92

CLAVARIA
No 5543
Clavaria stricta Pers
(Strict branch/upright coral)
North Chagrin Reservation
Willoughby Hills, Gates Mills, Mayfield, Ohio
20 August, 1942

Clavaria stricta has a cosmopolitan distribution, growing on dead wood, stumps, and branches of both leafy and coniferous trees.

CLAVARIA
No 6386
Clavaria stricta Pers
(Upright coral)
North Chagrin Reservation
Willoughby Hills, Gates Mills, Mayfield, Ohio
10 July, 1937

The *Clavaria stricta* is a tan or light brown color, though immediately bruises a light reddish-brown when handled. It is generally considered inedible.

94

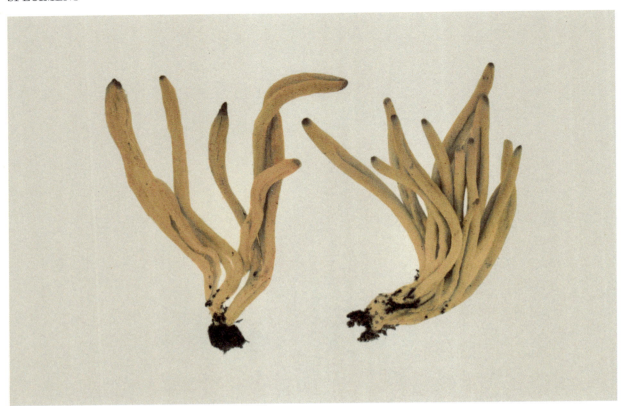

CLAVARIA
No 5535
Clavaria fumosa Pers
(Grayish fairy club, smoky clavaria, smoky spindles)
North Chagrin Reservation
Willoughby Hills, Gates Mills, Mayfield, Ohio
7 July, 1942

Walters noted that this study "should have a dash more of flesh color."

Clavaria fumosa is a type of saprobic fungus that grows in the soil of unfertilized grasslands, most commonly in clusters. The fungi's generic name comes from the Latin *clava*, meaning "club," and *fumosa*, meaning "smoky." Reportedly edible, the smoky clavaria is not commonly eaten due to its small stature (growing just 2–12 centimeters in height).

CLAVARIA
No 5544
Clavaria vermiculata (Micheli)
North Chagrin Reservation
Gates Mills, Ohio
24 June, 1942

(Top left) TREMELLA, No 6200. *Tremella reticulata* (Berk) Farlow (white coral jelly fungus). Holden Arboretum, Kirtland, Ohio. 10 September, 1940. *(Top right)* TREMELLODENDRON, No 6201. *Tremellodendron candidum (Schw) Atk.* North Chagrin Reservation, Gates Mills, Ohio. 6 June, 1939. *(Bottom)* TREMELLODENDRON, No 6201. *Tremellodendron candidum (Schw) Atk.* North Chagrin Reservation, Gates Mills, Ohio. 6 June, 1939.

The genus *Tremella* exists within the Tremellaceae family, a cosmopolitan group that contains both teleomorphic and anamorphic species, the latter of which are mostly yeasts. All *Tremella* species are parasites of other fungi—they are often found on wood-rotting fungi such as corticioid fungi—and derive their colloquial name, jelly fungi, from their distinctive gelatinous fruiting bodies. The genus was described by Carl Linneaus in his seminal text *Species Plantarum* (The Species of Plants) (1753), which classified every known plant at the time into genera and was the first publication to uniformly apply the structure of binomial nomenclature. Linnaeus named the genus for the Latin word *tremere,* meaning "to tremble." *Tremella fuciformis* (most commonly known as snow fungus) has been commercially cultivated for use in various cuisines, as well as in cosmetics and medicine, particularly in China and other Asian countries, including Korea, Vietnam, and Japan. *T. fuciformis* can be found in both tropical and subtropical climates, often appearing on species of hardwoods after heavy rainfall.

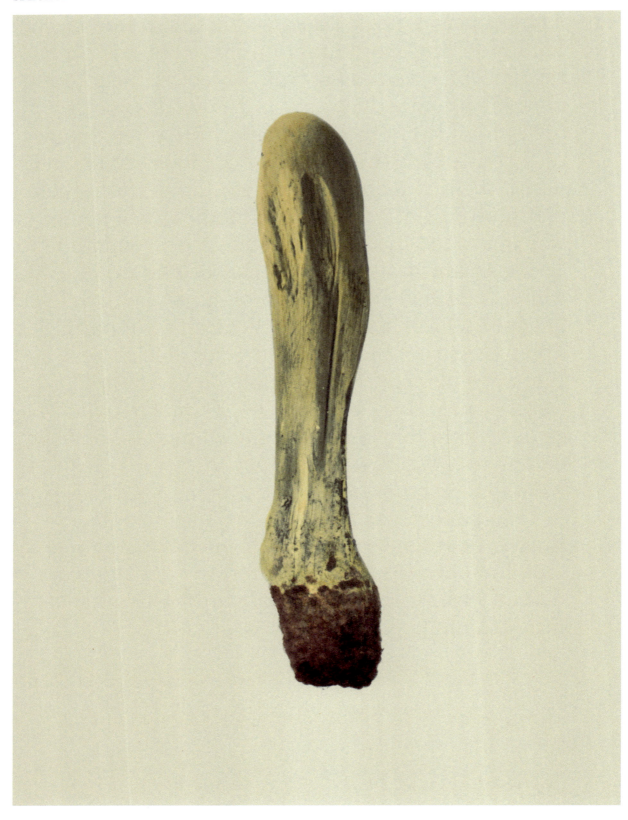

CLAVARIA
No 5541
Clavaria pistillaris L.
(Common club coral)
Holden Arboretum
Kirtland, Ohio
7 July, 1937

Clavaria pistillaris is a rare form of mushroom, commonly known as the common club coral, with a recognizable wrinkled fruiting body in the shape of a club. While the mushroom is edible, field guides have noted that it typically does not have enough flesh to make consumption worthwhile.

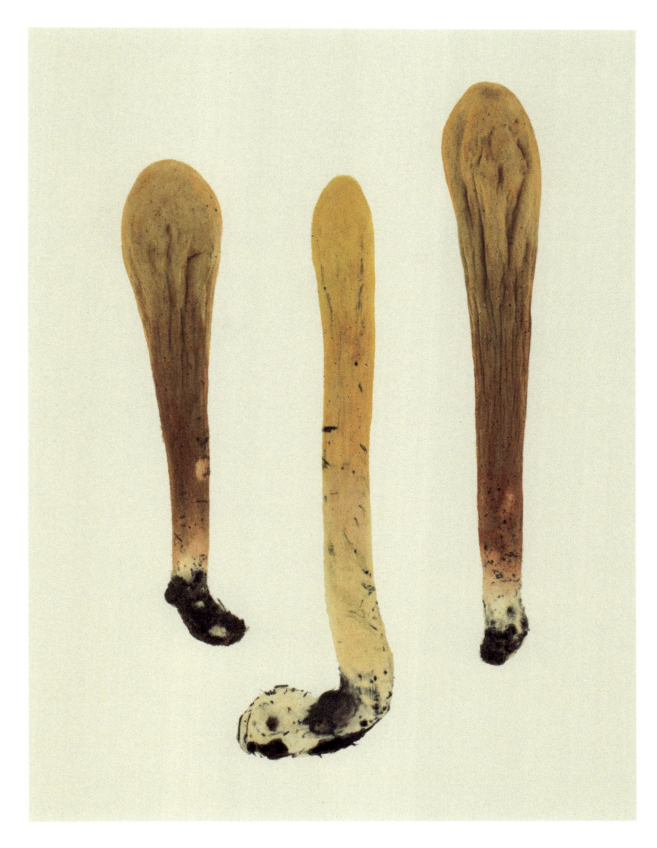

CLAVARIA
No 6347
Clavaria pistillaris L.
(Common club coral)
North Chagrin Reservation
Willoughby Hills, Gates Mills, Mayfield, Ohio
3 July, 1942

CLITOCYBE
No 5546
Clitocybe albidula Pk
North Chagrin Reservation
Willoughby Hills, Gates Mills, Mayfield, Ohio
18 August, 1942

MISC
No 6228
Volvaria bombycina Fr
Holden Arboretum
Kirtland, Ohio
7 August, 1941

CLITOCYBE
No 5553
Clitocybe cyathiformis Fr
(Goblet mushroom)
North Chagrin Reservation
Willoughby Hills, Gates Mills, Mayfield, Ohio
27 May, 1945

Growing in woodland across Europe, Asia, and North America, the *Clitocybe cyathiformis* has a dark, funnel-shaped cap and long, scaly stems. Its common name, goblet mushroom, references its conical cap shape.

CLITOCYBE
No 5568
Clitocybe monadelpha
North Chagrin Reservation
Willoughby Hills, Gates Mills, Mayfield, Ohio
Date unknown

CANTHARELLUS
No 5518
Cantharellus cibarius Fr (a six-spored specimen)
(Golden chanterelle)
North Chagrin Reservation
Willoughby Hills, Gates Mills, Mayfield, Ohio
14 July, 1937

In his notes for this specimen, Walters referred to a Mr. Beardslee who "thinks this is more apt to have been C. clavatus, or closely related to it." *Coccomyces clavatus* is a species of foliicolous fungus native to New Zealand.

CLITOCYBE
No 5562
Clitocybe laccata Fr
（Deceiver, lackluster laccaria）
North Chagrin Reservation
Willoughby Hills, Gates Mills, Mayfield, Ohio
16 June, 1939

The common names of *Clitocybe laccata* include deceiver and lackluster laccaria due to its highly variable appearance, with shades ranging from brick-red to salmon pink and brown. The fungi is edible and tastes mild, and is traditionally eaten by the Zapotec people of Oaxaca, Mexico.

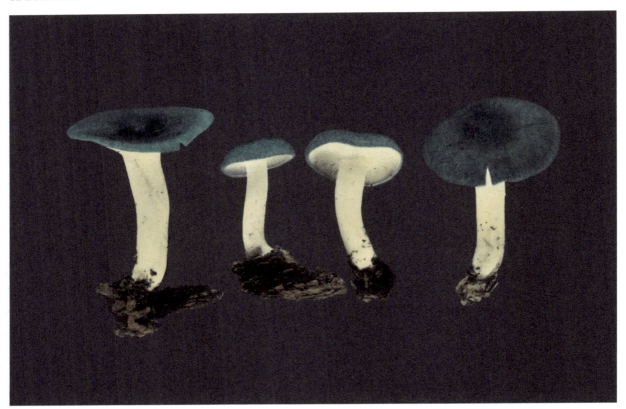

CLITOCYBE
No 5575
Clitocybe odora Fr var. anisearia
(Blue-green anise mushroom, aniseed toadstool)
Holden Arboretum
Kirtland, Ohio
8 September, 1940

The common name for *Clitocybe odora* is blue-green anise mushroom or aniseed toadstool, so-called for their strong anise odor. The mushroom is found in small groups among tree roots in deciduous and coniferous woodlands. Their caps are light blue when young, later fading to gray. The aniseed toadstool is edible, but similar in appearance to other poisonous fungi, so caution is advised in foraging.

CLITOCYBE
No 5563
Clitocybe laccata Fr var. Amethystina Bolt
North Chagrin Reservation
Gates Mills, Ohio
19 June, 1939

Walters noted that the coloring on this study was off, and that he should 'Change cap to same as gills, with over wash of 15-A-10; stems should be nearly 15-A-10."

COLLYBIA
No 5584
Collybia abundans Pk
Holden Arboretum
Kirtland, Ohio
6 August, 1942

108

(*Top*) MARASMIUS, No 5902. *Marasmius fagineus Morg.* North Chagrin Reservation, Willoughby Hills, Gates Mills, Mayfield, Ohio. 12 June, 1940. (*Bottom*) MISC, No 5855. *Lentinus cochleatus Fr* (aniseed cockleshell mushroom). North Chagrin Reservation, Willoughby Hills, Gates Mills, Mayfield, Ohio. 14 November, 1939.

(Top) COLLYBIA, No 5606. *Collybia succosa Pk.* North Chagrin Reservation, Willoughby Hills, Gates Mills, Mayfield, Ohio. 13 July, 1946. *(Bottom)* MISC, No 5899. *Marasmius delectans Morg.* North Chagrin Reservation, Gates Mills, Ohio. 31 July, 1942.

MISC
No 5987
Omphalia fibuloides Pk
Location unknown
28 May, 1937

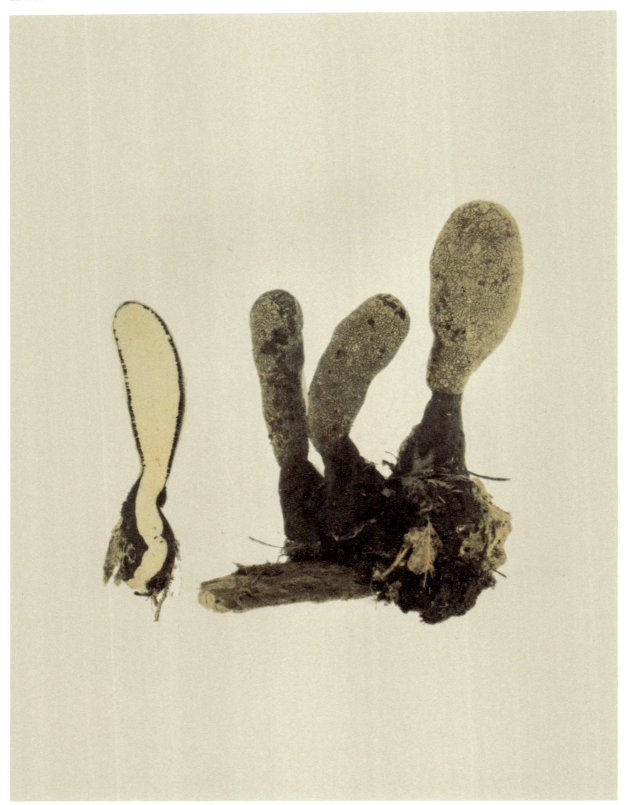

XYLARIA
No 6380
Xylaria polymorpha Pers
(Dead man's fingers)
North Chagrin Reservation
Willoughby Hills, Gates Mills, Mayfield, Ohio
1 August, 1943

112

Dead man's fingers (*Xylaria polymorpha*) is a cosmopolitan saprobic fungus—one that engages in lysotrophic nutrition, whereby chemoheterotrophic extracellular digestion is utilized in the processing of decayed (dead or waste) organic matter—characterized by elongated clavate, or strap-like, stromata, that poke up through the ground like fingers. As the specific epithet *polymorpha* ("many forms") suggests, there are variations in the fruiting body, though it most often takes a club-shaped form that has been likened to burned wood. Dead man's fingers belong to the phylum Ascomycetes, the sac fungi, characterized by their sac-like structure, which are usually found growing at the base of rotting or injured tree stumps and on decaying wood, or on wood objects that are in contact with soil. The basionym (original scientific name) of *Sphaeria polymorpha* was given to this ascomycetous fungus in 1797 by the Cape Colony mycologist Christiaan Hendrik Persoon, who is credited with making additions to Carl Linneaus's mushroom taxonomy.

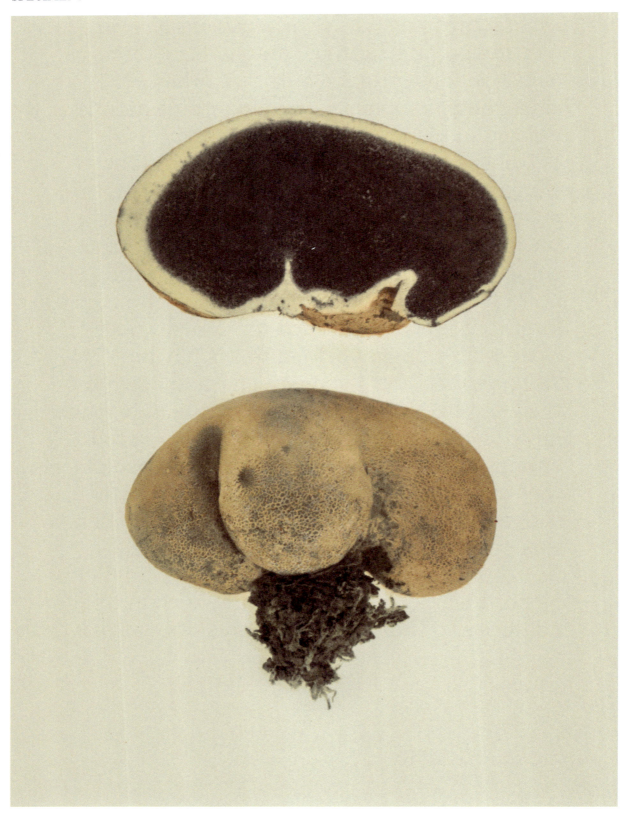

MISC
No 6412
Scleroderma lycoperdcides Schw
North Chagrin Reservation
Willoughby Hills, Gates Mills, Mayfield, Ohio
19 September, 1949

BOVISTA
No 5501
Bovista pila B & C
(Tumbling puffball)
Burton Wetlands Nature Preserve
Geauga County, Burton, Ohio
12 October, 1941

Bovista pila is commonly known as the tumbling puffball, as puffballs from this mushroom are known to detach and be blown by the wind. It is usually found in stables, open woods, and roadsides. Historically, various puffballs were used by the Indigenous Chippewa peoples of North America as a type of charm.

CALVATIA
No 5506
Calvatia craniformis (Schw) Fr
(Brain puffball, skull-shaped puffball)
North Chagrin Reservation
Willoughby Hills, Gates Mills, Mayfield, Ohio
4 November, 1941

The Latin *craniformis* roughly translates to "brain-shaped," referring to this puffball's resemblance to an animal brain. The puffball grows in open woods, hardwood forests, and wet areas in Asia, Australia, North America, and Mexico. It is an edible species, with young *Calvatia craniformis* having a mild odor and favorable taste. The brain puffball is used in Chinese and Japanese folk medicine.

Puffballs belong to the division Basidiomycota, which includes several genera such as *Calvatia*, *Calbovista*, and *Lycoperdon*. A polyphyletic assemblage—a group that includes organisms with mixed evolutionary origin but does not include their most recent common ancestor—all puffball fungi fruit into ball-shaped bodies that, when mature, will burst upon contact or impact and release a visible cloud of spores (giant puffballs, *Calvatia gigantea* can contain an estimated 7 trillion spores per specimen). With species found in locations ranging from North America to East Asia, puffballs have been used in a variety of contexts, including ink-making and beekeeping. In the US, people of the Lakota nation would use mature puffballs for wound care, utilizing their dry sacs of spores as coagulants. *C. gigantea* is now known to contain calvacin, a potent antitumor mucoprotein.

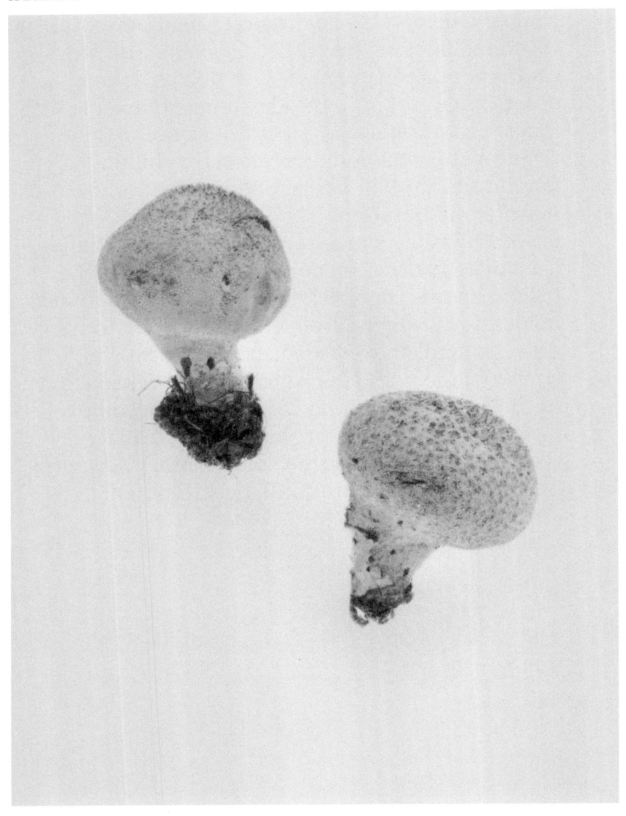

LYCOPERDON
No 5894
Lycoperdon pedicellatum Pk
North Chagrin Reservation
Willoughby Hills, Gates Mills, Mayfield, Ohio
3 September, 1944

MISC
No 6377
Scleroderma lycoperdoides Schw
North Chagrin Reservation
Willoughby Hills, Gates Mills, Mayfield, Ohio
5 September, 1943

119

ELVELA
No 6354
Elvela gigas Krombh
North Chagrin Reservation
Willoughby Hills, Gates Mills, Mayfield, Ohio
3 May, 1947

ELVELA
No 5691
Elvela underwoodii
North Chagrin Reservation
Willoughby Hills, Gates Mills, Mayfield, Ohio
Date unknown

Walters noted: "Log woods (spores warted and apiculate. Seaver's illustration shows spores lacking apiculi - immature, or another species). However, Plate 72 in his Supplement (N.A. Operculates) illustrates Elvela caroliniana with spores 25–30 × 12–14 and with knoblike apiculi, exactly like those of this specimen."

GEASTER
No 5732
Geaster saccatus Fr
（Rounded earthstar）
North Chagrin Reservation
Gates Mills, Ohio
31 July, 1942

The fruiting body of *Geaster saccatus* is initially egg-shaped, resembling a puffball mushroom, but then expands into a characteristic star shape, with rays that curve backward. The fungi is found across North, Central, and South America, as well as parts of Africa and Asia.

GEASTER
No 6355
Geaster triplex (Jungh)
(Collared earthstar, saucered earthstar)
North Chagrin Reservation
Willoughby Hills, Gates Mills, Mayfield, Ohio
26 September, 1950

Found in the detritus of hardwood forests across the world, the *Geaster triplex* is recognizable for its distinctive star shape.

Tough and fibrous, these fungi are not typically eaten, though there is no evidence to suggest inedibility. They have traditionally been used by Indigenous Americans for medicine; the Cherokee people would place the fruit bodies of *Geaster triplex* into the navel of a newborn until their withered umbilical cord fell off. The mushroom is also used in traditional Chinese medicine to reduce inflammation and swelling.

(Top) XYLARIA, No 6231. *Xylaria polymorpha Pers* (dead man's fingers). North Chagrin Reservation, Willoughby Hills, Gates Mills, Mayfield, Ohio. 11 June, 1940. *(Bottom)* COLLYBIA, No 5600. *Collybia plexipes Fr.* North Chagrin Reservation, Gates Mills, Ohio. 16 June, 1940.

HYPHOLOMA
No 5795
Hypholoma hydrophilum Fr
North Chagrin Reservation
Willoughby Hills, Gates Mills, Mayfield, Ohio
29 May, 1936

LEPIOTA
No 5883
Lepiota rhacodes (Vitt) Quel, Lepiota brunnea
(Farl & Burt)
(Shaggy/brown parasol)
Location unknown
6 September, 1942

126

LEPIOTA
No 5878
Lepiota morgani Pk
North Chagrin Reservation
Norwalk, Ohio
29 July, 1942

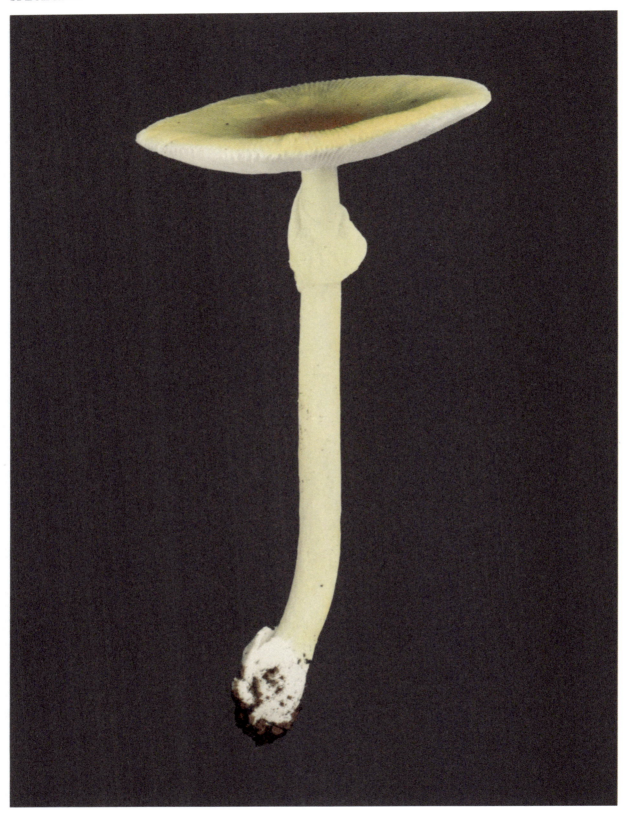

AMANITA
No 5400
Amanita frostiana Pk
North Chagrin Reservation
Gates Mills, Ohio
24 August, 1937

The *Amanita frostiana* has a recognizable cap that varies in color from yellow to scarlet and reddish-pink, though it is similar to other species in the *Amanita* genus, including *Amanita flavoconia* and *Amanita albocreata*. It is native to eastern North America and is typically found in oak forests.

AMANITA
No 5415
Amanita velatipes Atk
(Veiled bulb amanita)
North Chagrin Reservation
Gates Mills, Ohio
9 July, 1939

AMANITOPSIS
No 5429
Amanitopsis vaginata Fr
(Grisette)
North Chagrin Reservation
Gates Mills, Ohio
12 July, 1943

AMANITA
No 5412
Amanita spreta Pk
(Hated amanita)
North Chagrin Reservation
Willoughby Hills, Gates Mills, Mayfield, Ohio
7 July, 1942

AMANITA
No 5414
Amanita velatipes Atk
North Chagrin Reservation
Willoughby Hills, Gates Mills, Mayfield, Ohio
13 July, 1938

132

MISC
No 5426
Amanitopsis agglutinata, B & C
North Chagrin Reservation
Willoughby Hills, Gates Mills, Mayfield, Ohio
1 August, 1937

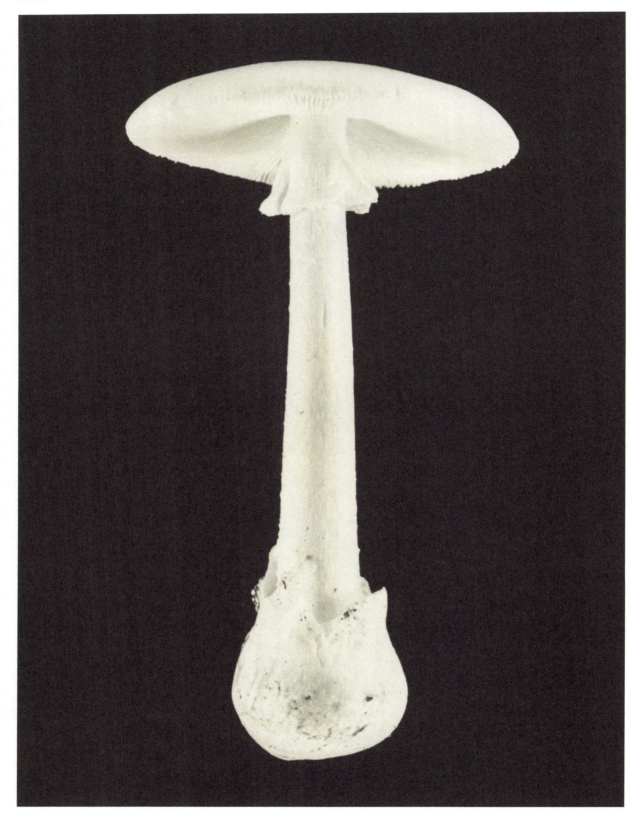

AMANITA
No 5423
Amanita verna Fr
North Chagrin Reservation
Willoughby Hills, Gates Mills, Mayfield, Ohio
18 September, 1940

134

The genus *Amanita* is perhaps the most famous in the known mycological universe. It contains approximately 600 species of agarics (fungal fruiting bodies where the pileus and stipe—cap and stalk—are clearly differentiated, with gills present on the underside of the cap), including *Amanitas verna, verosa,* and *bisporigera*—all of which are variants of the destroying angel. These all-white fungi are poisonous when consumed, largely due the presence of the potent amatoxin alpha-Amanitin. The genus is also home to one of the most recognizable species, *Amanita muscaria,* or fly agaric, which often grows in symbiosis with pine or birch trees. With its signature bright red cap and pyramid-shaped white warts (remnants of the universal veil membrane that enclosed the mushroom body when it was young), this basidiomycete occupies a distinct place in pop culture, as it is the visual inspiration for iconic elements of the platform game series *Super Mario Bros.* and the comic franchise *The Smurfs.*

AMANITA
No 6329
Amanita caesarea (Scop) Pers
(Caesar's mushroom)
Highlands, North Carolina
6 September, 1947

136

AMANITOPSIS
No 5427
Amanitopsis parcivolvata Pk
(Ringless false fly amanita)
Quarry Park North
South Euclid, Ohio
10 July, 1943

AMANITA
No 5396
Amanita flavoconia Atk
(Yellow patches, yellow wart, orange amanita)
North Chagrin Reservation
Willoughby Hills, Gates Mills, Mayfield, Ohio
26 July, 1947

This mushroom's specific epiphet *flavoconia* is the Latin for "yellowish," with its common names including yellow patches and yellow wart. It grows either alone or in groups on the ground between summer and fall in hemlock and red spruce forests across eastern North America.

AMANITA
No 5388
Amanita caesarea (Scop) Pers
North Chagrin Reservation
Willoughby Hills, Gates Mills, Mayfield, Ohio
7 July, 1942

AMANITA
No 5421
Amanita velatipes
(Veiled bulb amanita)
North Chagrin Reservation
Gates Mills, Ohio
1935 – 1951

AMANITA
No 5416
Amanita velatipes Atk
（Veiled bulb amanita）
North Chagrin Reservation
Willoughby Hills, Gates Mills, Mayfield, Ohio
26 June, 1942

AMANITA
No 5413
Amanita velatipes Atk
(Veiled bulb amanita)
North Chagrin Reservation
Willoughby Hills, Gates Mills, Mayfield, Ohio
26 July, 1935

The *Amanita velatipes* has a brownish center with many white warts, and typically grows among hardwood and conifer woods in summer and fall across the northeastern United States.

142

AMANITA
No. 5389
Amanita chlorinosma Pk
(Chlorine amanita)
North Chagrin Reservation
Willoughby Hills, Gates Mills, Mayfield, Ohio
9 September, 1940

The *Amanita chlorinosma* is considered highly toxic, and is recognizable by its strong odor of chlorine or rotting meat, its large size, and its enlarged base.

ENTOLOMA
No 5700
Entoloma salmoneum Pk
(Unicorn mushroom)
North Chagrin Reservation
Gates Mills, Ohio
9 July, 1939

Young *Entoloma salmoneum*s are a bright salmon orange color, though this fades with age. It is recognizable by its gnome-hat-like caps, and has the common name of unicorn mushroom for its fantastical appearance.

144

ENTOLOMA
No 5695
Entoloma cuspidatum Pk
Holden Arboretum
Kirtland, Ohio
11 July, 1938

HYGROPHORUS
No 5789
Hygrophorus puniceus Fr
(Crimson/scarlet waxcap)
Holden Arboretum
Kirtland, Ohio
10 October, 1938

The latter part of this species's generic name *Hygrophorus puniceus* translates to "blood red" in Latin, and its common names, crimson or scarlet waxcap, also highlight the fungi's distinctive color. This species has been marked as vulnerable on the International Union for Conservation of Nature (IUCN)'s Red List of Threatened Species.

146

HYGROPHORUS
No 5781
Hygrophorus miniatus Fr var. cantharellus
(Vermillion waxcap, miniature waxy cap)
North Chagrin Reservation
Willoughby Hills, Gates Mills, Mayfield, Ohio
2 July, 1938

The *Hygrophorus miniatus* has a distinctive bright red cap and stem, with its specific epiphet *miniata* from the Latin *miniat*, meaning "painted with red lead." A cosmopolitan species, the vermillion waxcap has been recorded across Europe and the United States, as well as Australia.

LACTARIUS
No 5829
Lactarius hygrophoroides B & C
North Chagrin Reservation
Willoughby Hills, Gates Mills, Mayfield, Ohio
18 August, 1935

148

ENTOLOMA
No 5702
Entoloma strictius Pk
North Chagrin Reservation
Willoughby Hills, Gates Mills, Mayfield, Ohio
11 August, 1938

PLUTEUS
No 6081
Pluteus flavofuligineus
North Chagrin Reservation
Gates Mills, Ohio
Date unknown

The deer shield (*Pluteus cervinus*), also known as the fawn mushroom, is a species of fungus in the order Agaricales. A saprotrophic species (one that uses chemoheterotrophic extracellular digestion to process decayed—dead or waste—organic matter), deer shields are often found on various kinds of wood waste, such as rotten logs, roots, and tree stumps. The specific epithet *cervinus* translates to "deer-like," a reference to the cap's brown color, first described by the German mycologist Jacob Christian Schäffer in 1774 as *rehfarbig,* meaning "fawn-colored."

BOLETUS
No 5448
Boletinus spectabilis
Burton Wetlands Nature Preserve
Geauga County, Burton, Ohio
24 September, 1944

BOLETUS
No 5450
Boletinus spectabilis
Burton Wetlands Nature Preserve
Geauga County, Burton, Ohio
24 September, 1944

BOLETUS
No 5473
Boletus fraternus
North Chagrin Reservation
Gates Mills, Ohio
Date unknown

154

POLYPORUS
No 6099
Polyporus cinnabarinus (Jacq) Fr
(Cinnabar polypore)
North Chagrin Reservation
Gates Mills, Ohio
16 June, 1939

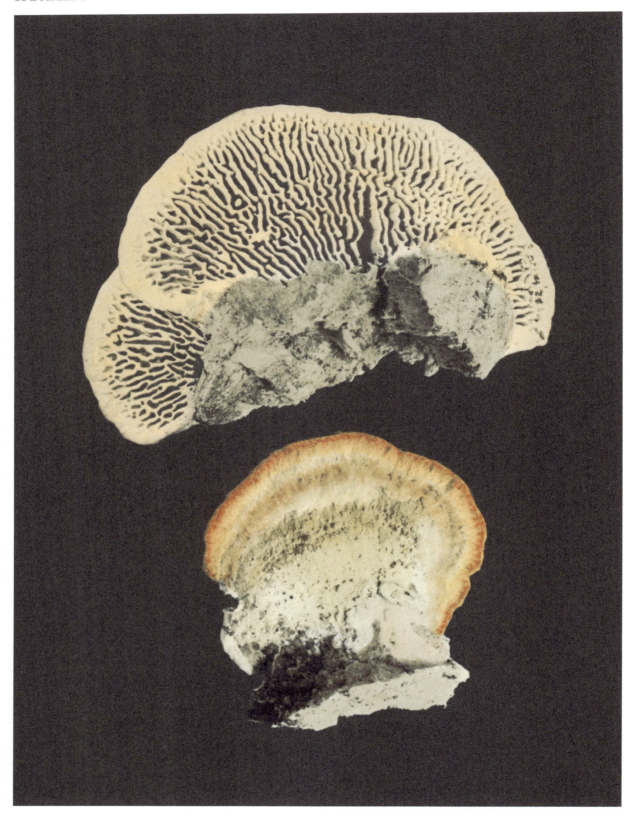

DAEDALEA
No 5671
Daedalea quercina (L) Fr
Location unknown
29 September, 1938

156

Daedalea is a genus of fungi in the Fomitopsidaceae family, which includes species such as *Fomes fomentarius* (the Iceman fungus). Its generic name derives from the ancient Greek for "curiously wrought," a reference to the patterns created by the maze-like pores on the underside of the fruiting bodies of the fungi in this genus. Cape Colony mycologist Christian Hendrik Persoon was the first to circumscribe—the delimitation of which subordinate taxa are parts of a specific taxon—the genus in 1801, based on the type species *Daedalea quercina,* which, in 2024, was transferred to the genus *Fomitopsis,* making it *Fomitopsis quercina.* The *Fomitopsis* genus was first circumscribed by the Finnish mycologist Petter Karsten in 1881, who based the type species off of *Fomitopsis pinicola,* the red-belted conk.

POLYPORUS
No 6117
Polyporus hirsutus (Wulf) Fr
(Hairy bracket, hairy turkey tail)
North Chagrin Reservation
Willoughby Hills, Gates Mills, Mayfield, Ohio
11 June, 1939

This polyporus is found on the dead wood of trees, particularly beechwood, and is very common in Britain and Northern Ireland, but can also be found in China and North America.

STEREUM
No 6188
Stereum rameale Schw
North Chagrin Reservation
Willoughby Hills, Gates Mills, Mayfield, Ohio
29 August, 1939

The *Stereum rameale* is a form of plant fungus that infects trees, most commonly peach but also beech trees. *Stereum* means "tough," which this species is particularly when dry. It is a tasteless and inedible fungi.

AMANITA
No 5390
Amanita chlorinosma Pk
North Chagrin Reservation
Willoughby Hills, Gates Mills, Mayfield, Ohio
20 July, 1941

AMANITA
No 5391
Amanita chlorinosma Pk
North Chagrin Reservation
Willoughby Hills, Gates Mills, Mayfield, Ohio
25 August, 1942

AMANITA
No 5397
Amanita flavo-rubescens
Location unknown
Date unknown

Amanita flavo-rubescens are mycorrhizal with hardwoods, particularly oaks, and are found throughout North America and Mexico. They are very similar to *Amanita rubescens,* but are differentiated by their yellow cap and flaky yellow warts. Similarly to the blushing amanita, *Amanita flavo-rubescens* stains a dark reddish color.

ARMILLARIA
No 5438
Armillaria granosa (Morg)
Holden Arboretum
Kirtland, Ohio
29 September, 1937

AMANITA
No 5410
Amanita rubescens Fr
(Blushing amanita),
North Chagrin Reservation
Willoughby Hills, Gates Mills, Mayfield, Ohio
26 June, 1942

The *Amanita rubescens* is commonly known as the blushing amanita due to the tendency of its flesh to turn pink after being cut. It is an edible mushroom popular in many European countries, though there is evidence it can cause anemia if eaten in large quantities.

164

AMANITA
No 5430
Amanita vaginata Fr var. Fulva
(Grisette)
North Chagrin Reservation
Willoughby Hills, Gates Mills, Mayfield, Ohio
5 August, 1943

The *Amanita vaginata* or grisette is widespread in North America and Europe, growing in coniferous or hardwood forests. While edible, it is strongly advised not to consume grisette mushrooms due to their resemblance to poisonous species of *Amanita* such as the death cap (*Amanita phalloides*).

PANAEOLUS
No 5991
Panaeolus subbalteatus Berk
(Banded mottlegill, weed panaeolus, belted panaeolus)
Lake View Cemetery
Cleveland, Ohio
2 June, 1943

Walters noted that he found this specimen in a dung heap.

Panaeolus subbalteatus, often known as the belted panaelous or the banded mottlegill, occurs on dung or dung-rich soil, tending to grow in small clusters. The mushroom is mildly psychoactive, and is one of the most common psilocybin mushrooms found in the United States.

AMANITA
No 5385
Amanita caesarea Fr
(Caesar's mushroom)
Willoughby Hills, Gates Mills, Mayfield, Ohio
2 July, 1940

This mushroom's common name, coined in 1772 by Giovanni Antonio Scopoli, is derived from *Amanita caesarea* being a favorite of early rulers of the Roman Empire, particularly emperor Claudius.

Caesar's mushrooms are edible but look similar to the deadly poisonous death cap (*Amanita phalloides*) and destroying angels mushrooms, so should be identified with absolute certainty before consuming.

AMANITA
No 6394
Amanita vaginata Fr.
(Grisette)
Brecksville Reservation
Brecksville, Ohio
12 July, 1950

The *Amanita vaginata* or grisette is widespread in North America
and Europe, growing in coniferous or hardwood forests. Whilst edible,
it is strongly advised not to consume grisette mushrooms due to their
resemblance of poisonous species of *Amanita* such as the death cap
(*Amanita phalloides*).

168

AMANITA
No 5406
Amanita porphyria
(Gray veiled amanita)
North Chagrin Reservation
Willoughby Hills, Gates Mills, Mayfield, Ohio
11 June, 1947

Amanita porphyria, known commonly as the gray veiled amanita, is not suitable for consumption and can be easily confused with the highly toxic death cap mushroom (*Amanita phalloides*).

BOLETUS
No 5452
Boletus
North Chagrin Reservation
Gates Mills, Ohio
12 July, 1943

170

BOLETUS
No 5455
Boletus albidus subsp. eupachypus
(Rooting bolete)
Quarry Park North
South Euclid, Ohio
9 August, 1945

When young, *Boletus albidus* has an off-white to gray cap that darkens with age, and yellow, sponge-like pores that bruise blue. It is not poisonous, but has an incredibly bitter taste.

MISC
No 6333
Boletinus cavipes (opat.) Kalchbr
Geauga County, Ohio
16 October, 1951

BOLETUS
No 6385
Boletus betula Schw
（Shaggy stalked bolete)
Highlands, North Carolina
4 September, 1947

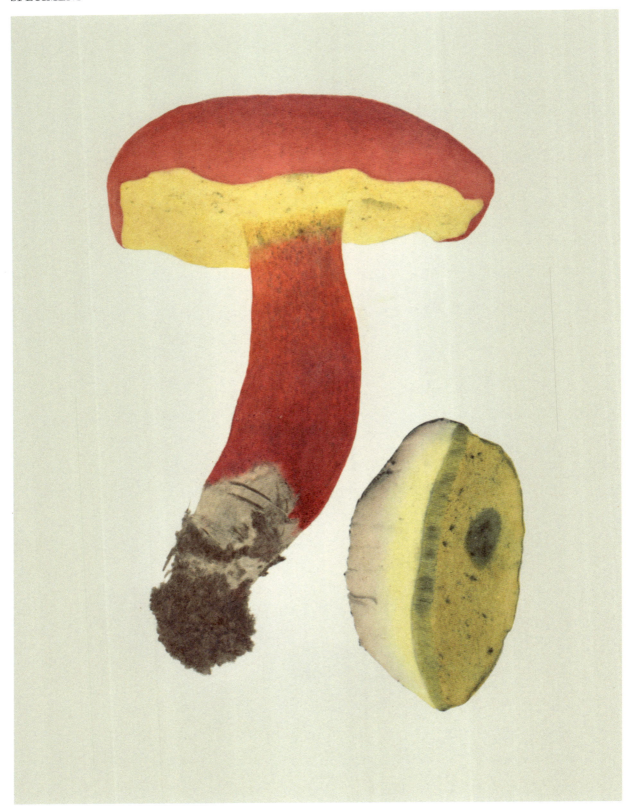

BOLETUS
No 5459
Boletus bicolor Pk
(Baorangia bicolor)
Gates Mills, Ohio
24 July, 1943

Known for its two-tone color scheme, the *Baorangia bicolor*, a name coined in 1807 by Italian botanist Giuseppe Raddi, is an edible fungus found in eastern North America, as well as China and Nepal.

The cap's varying color can be used to determine if the mushroom is ready to be eaten; a lighter red cap means the mushroom is less mature, with a dark brick-red cap indicating it is safe to eat.

BOLETUS
No 5474
Boletus frostii Russell (Boletus alveolatus B & C)
(Frost's bolete, apple bolete)
North Chagrin Reservation
Gates Mills, Ohio
18 August, 1937

This species was named in the late 19th century by the Unitarian minister John Lewis Russell of Salem, Massachusetts, after his friend and fellow mycologist Charles Christopher Frost. *Boletus frostii Russell* is commonly known as Frost's bolete or apple bolete, and in Mexico is referred to as *panza agria*, which translates to "sour belly."

(Top) BOLETUS, No 5460. *Boletinus cavipes (opat.) Kalchbr* (hollow foot). Burton Wetlands Nature Preserve, Geauga County, Burton, Ohio. Date unknown. *(Bottom)* BOLETUS, No 5479. *Boletus indecisus* (indecisive bolete). North Chagrin Reservation, Gates Mills, Ohio. Date unknown.

(Top) BOLETUS, No 5480. *Boletus luteus L* (slippery jack, sticky bun). Holden Arboretum, Kirtland, Ohio. 24 October, 1939. *(Bottom)* BOLETUS, No 5487. *Boletus parasiticus Bull.* Burton Wetlands Nature Preserve, Geauga County, Burton, Ohio. Date unknown.

BOLETUS
No 5483
Boletus miniato-olivaccus Fr (faded)
Holden Arboretum
Kirtland, Ohio
24 August, 1944

The *Boletus miniato-olivaccus* has a white to pale yellow cap with a thin red layer under the cap skin. The cap stains blue and has a bitter taste.

178

BOLETUS
No 5498
Boletus subrelutipes Pk
（Red-mouth bolete）
North Chagrin Reservation
Gates Mills, Ohio
Date unknown

The cap of *Boletus subrelutipes* ranges from reddish-brown to reddish-orange, and its flesh instantly stains blue once cut, later becoming white. The red-mouth bolete is associated with deciduous trees, particularly oak and pine. This mushroom is toxic and can cause gastrointestinal problems if consumed.

BOLETUS
No 5490
Boletus scaber Bull
North Chagrin Reservation
Gates Mills, Ohio
26 July, 1937

180

BOLETUS
No 6341
Boletus indecisus Pk
(Indecisive bolete)
North Chagrin Reservation
Willoughby Hills, Gates Mills, Mayfield, Ohio
26 July, 1937

ENTOLOMA
No 5696
Entoloma cuspidatum Pk
North Chagrin Reservation
Willoughby Hills, Gates Mills, Mayfield, Ohio
26 June, 1942

182

CANTHARELLUS
No 5523
Cantharellus umbonatus Fr (in Polytrichum ohioense)
(Ohio hair-cap moss)
Gates Mills, Ohio
26 September, 1945

184

CLAVARIA
No 5528
Clavaria abietina Pers
(Green-staining coral)
North Chagrin Reservation
Willoughby Hills, Gates Mills, Mayfield, Ohio
27 September, 1945

Lepiota rhacodes typically grows in rings, resulting in the mushroom taking on a mythical quality, historically associated in European folklore with a place in the woods where fairies, elves, and pixies danced. Many cultures warned against humans entering these rings as they would die young or be transported to the fairy world. In reality, the rings are a natural occurrence caused by the way mycelium grows underground, outward and in a circular fashion in search of nutrients.

CLAVARIA
No 5533
Clavaria cristata (Holmsk) Pers
(Crested coral)
Quarry Park North
Euclid, Ohio
10 July, 1943

(Top) CLAVARIA, No 5529. *Clavaria amethystina (Batt) Pers* (coral fungus). North Chagrin Reservation, Gates Mills, Ohio. 19 June, 1939. *(Bottom)* CLAVARIA, No 5534. *Clavaria formosa* (pinkish coral mushroom, salmon coral, beautiful clavaria). Kirtland, Ohio. Date unknown.

CLAVARIA
No 5532
Clavaria cristata (Holmsk) Pers
(Crested coral)
Gates Mills, Ohio
19 June, 1939

Clavaria cristata, known as crested coral, is most commonly found in British, Irish, and North American woodlands, usually in wet areas. Other coral fungi, including *Clavulina rugosa* and *Clavulina cinerea*, have similar features to *Clavaria cristata*, which can create problems in identification.

MISC
No 5628
Cordyceps militaris
(Caterpillar fungus)
Station touristique Duchesnay
Sainte-Catherine-de-la-Jacques-Cartier, Québec
Date unknown

The *Cordyceps militaris,* known commonly as caterpillar fungus, can grow up to 8 centimeters in distinctive club-shaped red-and-orange fruiting bodies. Though considered inedible by North American foraging guides, the fungi is used in Asia in dishes including chicken soup and hot pot, as well as in traditional Chinese medicine.

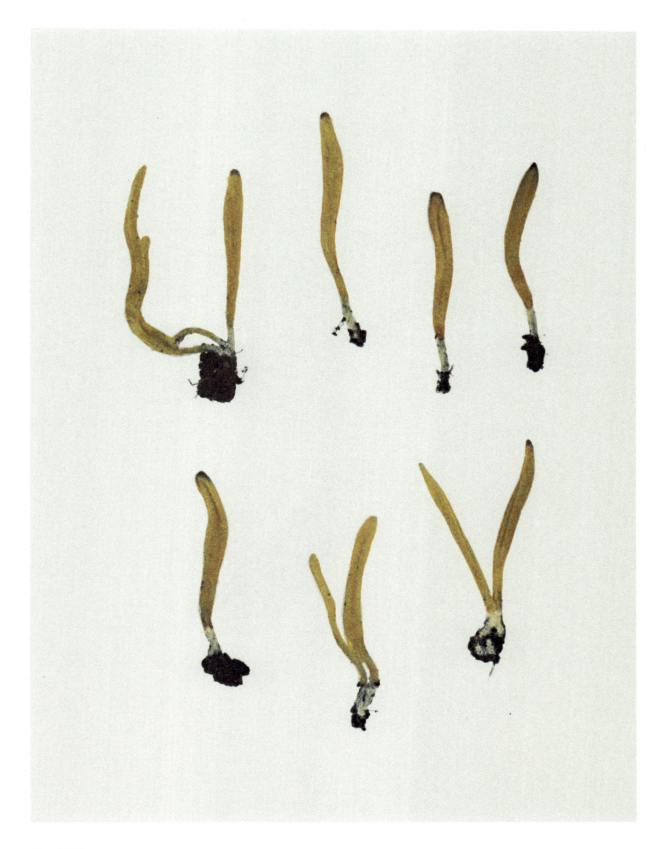

CLAVARIA
No 5542
Clavaria pulchra Pk
North Chagrin Reservation
Willoughby Hills, Gates Mills, Mayfield, Ohio
11 August, 1942

CLITOCYBE
No 5580
Clitocybe pulcherrima Pk
North Chagrin Reservation
Willoughby Hills, Gates Mills, Mayfield, Ohio
14 October, 1937

190

CLITOCYBE
No 5557
Clitocybe gigantea (Fr) Quel
(Giant clitocybe)
North Chagrin Reservation
Willoughby Hills, Gates Mills, Mayfield, Ohio
20 August, 1942

An extremely large mushroom, the *Clitocybe gigantea* is a saprobic species whose cap can reach a diameter of up to 50 centimeters. It often forms in fairy rings in pastures and on roadsides. The giant clitocybe is not poisonous, but has a smell that has been compared to fish meal.

COLLYBIA
No 5588
Collybia cirrata Fr
Thompson Ledges Township Park
Thompson, Ohio
29 September, 1938

The *Collybia cirrata* is a saprobic mushroom, typically found on the decaying remains of other mushrooms. It is widespread in Europe and North America, as well as Korea and Japan.

COLLYBIA
No 5589
Collybia colorea Pk
North Chagrin Reservation
Willoughby Hills, Gates Mills, Mayfield, Ohio
6 August, 1939

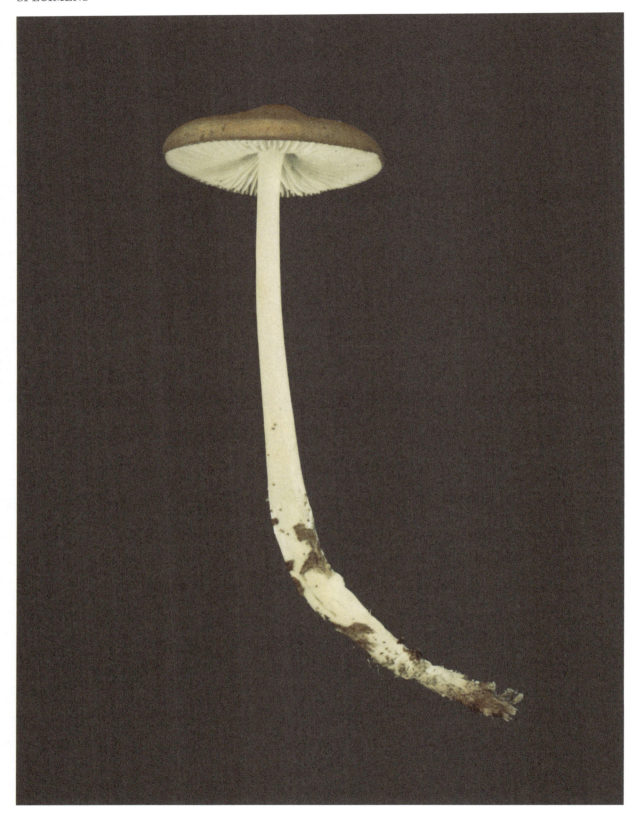

COLLYBIA
No 5588
Collybia cirrata Fr
Thompson Ledges Township Park
Thompson, Ohio
29 September, 1938

COLLYBIA
No 5589
Collybia colorea Pk
North Chagrin Reservation
Willoughby Hills, Gates Mills, Mayfield, Ohio
6 August, 1939

COPRINUS
No 5618
Coprinus comatus
(Shaggy ink cap, lawyer's wig, shaggy mane)
Cleveland Heights, Horseshoe Lake,
Shakers Lake, Ohio
Date unknown

When fully developed, the *Coprinus comatus* has bell-shaped white caps covered in scales. This mushroom is incredibly unusual, as it will turn black and dissolve a few hours after being picked. While edible, it must be eaten soon after picking, before its color changes.

COPRINUS
No 6401
Coprinus atramentarius Fr
(Common ink cap, tippler's bane, inky cap)
North Chagrin Reservation
Willoughby Hills, Gates Mills, Mayfield, Ohio
27 October, 1948

One of the common names for this mushroom, tippler's bane, is a reference to the potentially poisonous effects of consuming *Coprinus atramentarius* alongside alcohol, as it heightens the body's sensitivity to ethanol. Doing so leads to disulfiram syndrome, symptoms of which include nausea, vomiting, facial reddening, and malaise, which occurs mere minutes after consuming alcohol.

MISC
No 6381
Unidentified species
Location unknown
Date unknown

CORTINARIUS
No 5631
Cortinarius annulatus
North Chagrin Reservation
Willoughby Hills, Gates Mills, Mayfield, Ohio
29 August 1942

DAEDALEA
No 5669
Daedalea confragosa (Bolt) ex Fr
(Thin-walled maze polypore, blushing bracket)
Holden Arboretum
Kirtland, Ohio
29 September, 1937

Daedalea confragosa is a plant pathogen that causes decay of sapwood and a form of rot called white rot. While inedible, the thin-walled maze polypore is used in ornamental papermaking, with the fruit bodies pulped, pressed, and dried to create paper.

CRATERELLUS
No 5648
Craterellus cornucopioides Fr
(Black chanterelle, black trumpet, trumpet of the dead)
North Chagrin Reservation
Willoughby Hills, Gates Mills, Mayfield, Ohio
18 August, 1935

The *Cratrellus cornucopioides* has a funnel-shaped cap with a black inner surface. It is found in woods across North America, Europe, Asia, and Australia. The black chanterelle is a popular edible mushroom, and when dried has notes of black truffle.

BOLETUS
No 5491
Boletus separans
Location unknown
Date unknown

Walters annotated this study with: "The Bitter Boletus. A color sometimes occurring in the early stages – One of my earliest. No notes. Identity guessed at by Mr. B. from photo only. More likely it was B. separans."

ALEURIA
No 5379
Aleuria aurantia (pers) Fuckel
(Orange peel fungus)
North Chagrin Reservation
Willoughby Hills, Gates Mills, Mayfield, Ohio
14 October, 1938

The common name of this mushroom is a reference to its resemblance to orange peels strewn on the ground. *Aleuria aurantia* is edible but hard to collect intact.

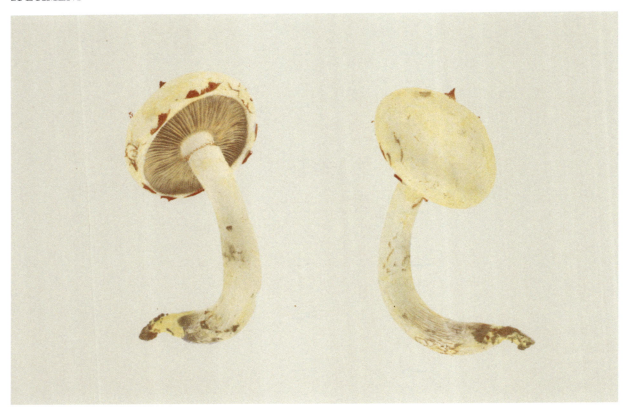

(Top) FLAMMULA, No 5710. *Flammula flavida (Fr)*. North Chagrin Reservation, Willoughby Hills, Gates Mills, Mayfield, Ohio. 27 October, 1946. *(Bottom)* FLAMMULA, No 5713. *Flammula lenta*. North Chagrin Reservation, Willoughby Hills, Gates Mills, Mayfield, Ohio. 1937–1941.

ENTOLOMA
No 5703
Entoloma strictius Pk
North Chagrin Reservation
Gates Mills, Ohio
12 July, 1943

FLAMMULA
No 5711
Flammula lenta Fr
North Chagrin Reservation
Willoughby Hills, Gates Mills, Mayfield, Ohio
15 October, 1937

206

FLAMMULA
No 5714
Flammula sapinea Fr
North Chagrin Reservation
Willoughby Hills, Gates Mills, Mayfield, Ohio
7 October, 1938

FLAMMULA
No 5716
Flammula spumosa
North Chagrin Reservation
Willoughby Hills, Gates Mills, Mayfield, Ohio
Date unknown

208

FLAMMULA
No 5715
Flammula spumosa Fr
North Chagrin Reservation
Willoughby Hills, Gates Mills, Mayfield, Ohio
5 September, 1943

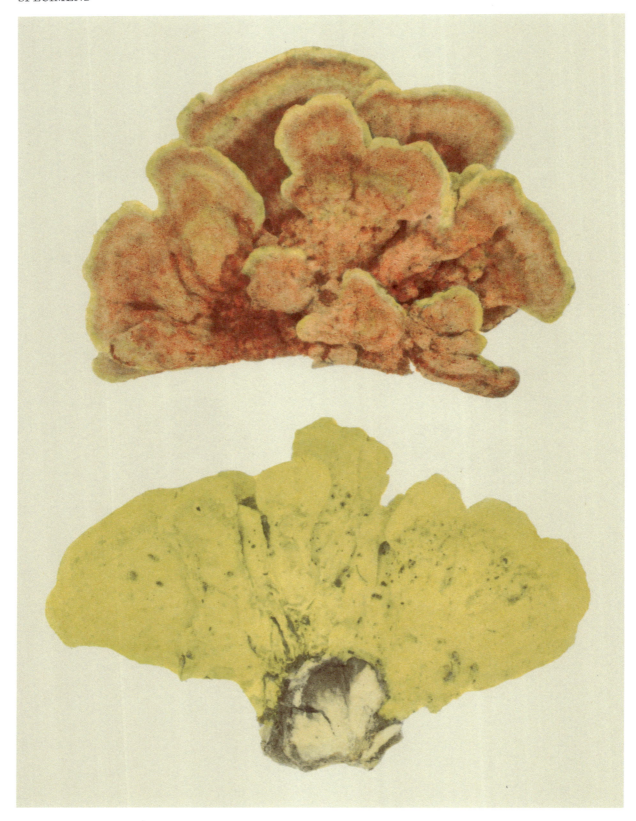

POLYPORUS
No 6374
Polyporus sulphureus Bull
North Chagrin Reservation
Willoughby Hills, Gates Mills, Mayfield, Ohio
11 September, 1945

210

Chicken-of-the-woods (*Laetiporus sulphureus*, also known as *Polyporus sulphureus Bull*), or sulphur shelf, is a species of bracket fungi that can be found growing on both living and dead hardwood (especially oak) and conifer trees. When young, the species is prized for its culinary uses; as its colloquial name suggests, it has a poultry-like taste and texture. French mycologist Pierre Bouillard was the first to describe the species in 1789. It was given its current name by the American mycologist William Murrill in 1920—the generic and specific epithets translate to "with bright pores" and "the color of sulfur," respectively. On October 15, 1990, Giovanni Paba discovered an example of *Laetiporus sulphureus* in the New Forest of Hampshire, United Kingdom, that weighed 100 lbs (45.35 kg). It was subsequently designated the "heaviest edible fungi" by Guinness World Records.

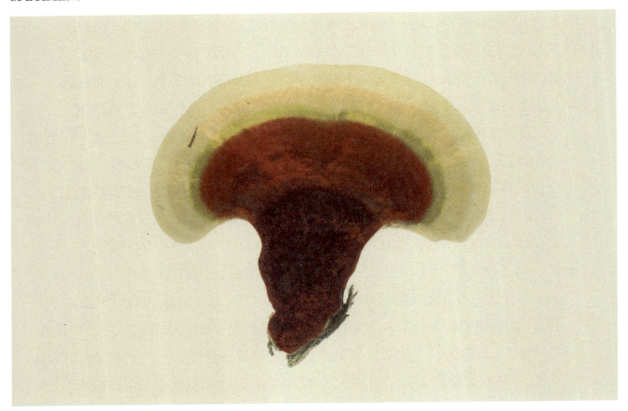

MISC
No 6404
Ganoderma lucidum (Leyss) Karst
North Chagrin Reservation
Willoughby Hills, Gates Mills, Mayfield, Ohio
3 June, 1941

POLYPORUS
No 6418
Polyporus graveolens (Schw) Fr
North Chagrin Reservation
Willoughby Hills, Gates Mills, Mayfield, Ohio
3 September, 1949

FOMES
No 5722
Fomes fomentarius (L) Gill
(Tinder fungus, hoof fungus, Iceman fungus)
Holden Arboretum
Kirtland, Ohio
28 July, 1939

214

The Iceman fungus (*Fomes fomentarius*) is a species of fungal plant pathogen. Its colloquial name derives from the four pieces of the fungus found in Ötzi the Iceman's possession when his Copper Age body was discovered in 1991 in the Ötzal Alps on the border between Austria and Italy. The spongy material amadou, a felt-like artificial leather, is derived from the fungus, and was utilized as a fire starter by ancient peoples (the specific epithet derives from the Latin *fomentum,* meaning "fuel")—a possible explanation of why Ötzi was carrying the fungus at the time of his death. *F. fomentarius* produces large polypore fruiting bodies (the undersides of which feature pores or tubes) that can measure between 5 and 45 centimeters in diameter. They are often likened to horse hooves, hence the species's other colloquial name, hoof fungus. With circumboreal distribution, they are commonly found growing on the sides of hardwood trees, their presence infecting the host species through damaged bark or broken branches and causing stem decay (rot). Once the host tree has died, *F. fomentarius* remains, transforming its status from that of parasite to decomposer.

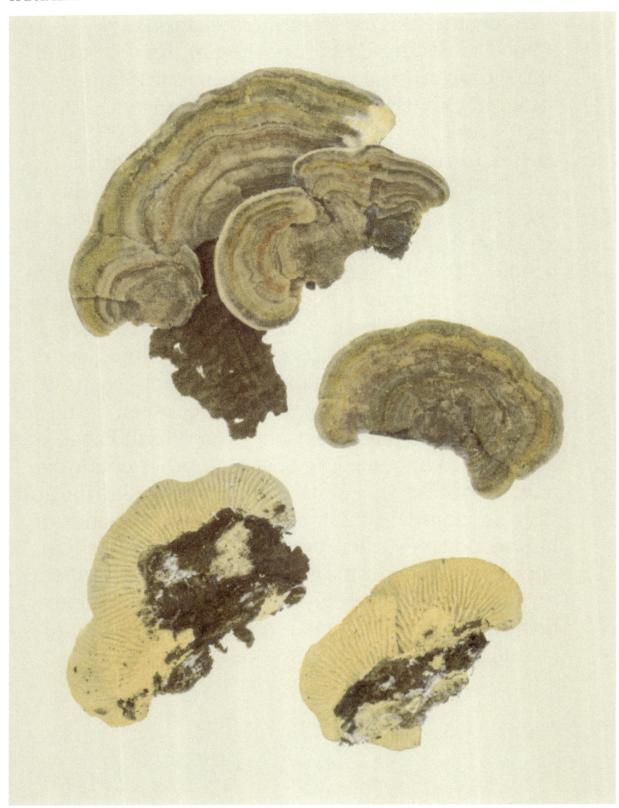

MISC
No 5858
Lenzites betulina L
Peninsula, Ohio
26 August, 1945

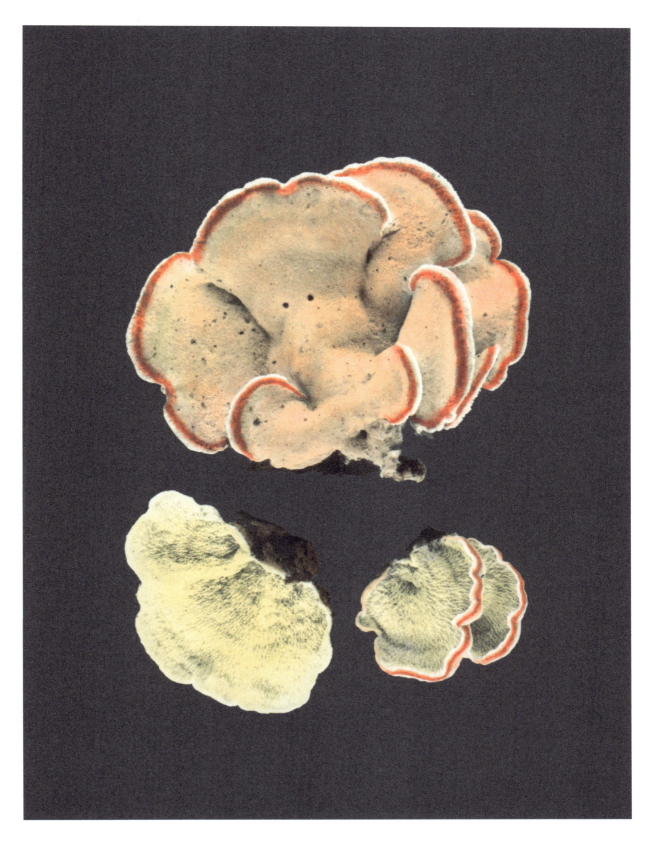

MISC
No 5917
Merulius rubellus Pk
North Chagrin Reservation
Willoughby Hills, Gates Mills, Mayfield, Ohio
11 September, 1939

(Top) MISC, No 5950. *Mycena fagicola (Smith)*. Holden Arboretum, Kirtland, Ohio. 2 November, 1941. *(Bottom)* MISC, No 5950. *Mycena fagicola (Smith)*. Holden Arboretum, Kirtland, Ohio. 2 November, 1941.

MISC
No 5947
Mycena cyaneobasis Pk
North Chagrin Reservation
Willoughby Hills, Gates Mills, Mayfield, Ohio
15 June, 1937

MISC
No 5957
Mycena niveipes (Murrill)
North Chagrin Reservation
Willoughby Hills, Gates Mills, Mayfield, Ohio
22 May, 1946

MISC
No 5912
Marasmius siccus Schw
North Chagrin Reservation
Willoughby Hills, Gates Mills, Mayfield, Ohio
31 August, 1935

GALERA
No 5725
Galera hypnorum Fr
(Moss bell)
Cleveland Heights, Ohio
17 June, 1941

GALERA
No 5728
Galera tenera Fr
（Conecap mushroom）
Cleveland Heights, Ohio
14 June, 1937

The *Galera tenera*, known commonly as the conecap mushroom, initially has a conical-shaped cap, becoming bell-shaped as it ages. Field guides have recorded this mushroom as poisonous.

GYMNOSPORANGIUM
No 6356
Gymnosporangium juniperi-virginianae
(Juniper apple rust)
North Chagrin Reservation
Willoughby Hills, Gates Mills, Mayfield, Ohio
13 May, 1945

224

Cedar-apple rust (*Gymnosporangium juniperi-virginianae*) is a plant pathogen that often develops its distinctive orange outcroppings on uniper stems after late-spring and summer rainstorms. These particular weather events cause the galls (abnormal growths on the external tissues of plants) that the fungi have formed on their host to swell and produce gelatinous forms known as telial horns, which can measure between 10 and 20 millimeters long. In order to develop and complete its life cycle, *C. juniperi-virginianae* requires that two hosts be present, often in relatively close proximity: a juniper species (such as eastern redcedar) and an apple species (such as crabapple, hawthorne, or quince). Where present, this rust fungus has the potential to cause widespread economic devastation, as it often affects crop plants and lumber trees. Since the infecting spores are carried on the wind, juniper trees are often cut down within a 2- to 3-mile radius of apple-growing areas. In Walters's photograph of cedar-apple rust, he has included instances of galls both with and without the presence of telial horns, as well as a full section of juniper foliage, which contextualizes the composition and the fungi's specific life cycle.

HYPHOLOMA
No 5794
Hypholoma ericeum (Fr) Kuhner
North Chagrin Reservation
Gates Mills, Ohio
26 September, 1945

Walters noted that this mushroom is usually found in beds of *Polytrichum*, a genus of mosses.

226

HYDNUM
No 6358
Hydnum auriscalpium L
(Pinecone mushroom, cone tooth, ear-pick fungus)
North Chagrin Reservation
Willoughby Hills, Gates Mills, Mayfield, Ohio
1 July, 1945

The specific epiphet *auriscalpium* is Latin for "ear pick," hence this mushroom's common name, ear-pick fungus, due to the fine brown hairs that cover the small caps of the species. A saprobic species, the pinecone mushroom tends to grow on fallen pine cones as well as spruce and Douglas fir cones.

(Top) HYGROPHORUS, No 5775. *Hydnum erinaceus (Bull)* (lion's mane, yamabushitake, beared tooth fungus, bearded hedgehog). North Chagrin Reservation, Willoughby Hills, Gates Mills, Mayfield, Ohio. 6 October, 1939.
(Bottom) HYGROPHORUS, No 5778. *Hygrophorus marginatus Pk*. North Chagrin Reservation, Willoughby Hills, Gates Mills, Mayfield, Ohio. 5 July, 1940.

(*Top*) HYGROPHORUS, No 5785. *Hygrophorus pratensis Fr* (meadow waxcap). Holden Arboretum, Kirtland, Ohio. 4 July, 1940. HYGROPHORUS, No 5788. *Hygrophorus pudorinus Fr* (blushing waxycap, turpentine waxycap). Holden Arboretum, Kirtland, Ohio. 8 October, 1940.

CORTINARIUS
No 6351
Cortinarius atkinsonianus Kauff
North Chagrin Reservation
Willoughby Hills, Gates Mills, Mayfield, Ohio
28 September, 1950

CORTINARIUS
No 5640
Cortinarius mucifluus Fr
North Chagrin Reservation
Gates Mills, Ohio
17 September, 1942

BOLETUS
No 5461
Boletus chromapes
(Yellow foot bolete)
State College, Pennsylvania
Date unknown

232

ARMILLARIA
No 5441
Armillaria mellea (Vahl) Fr
North Chagrin Reservation
Gates Mills, Ohio
21 September, 1935

ENTOLOMA
No 5697
Entoloma grayanum Pk
North Chagrin Reservation
Willoughby Hills, Gates Mills, Mayfield, Ohio
20 August, 1942

234

CLITOCYBE
No 5573
Clitocybe ochropurpurea Berk
North Chagrin Reservation
Gates Mills, Ohio
5 September, 1935

HYPHOLOMA
No 5799
Hypholoma radicosum lange
(Rooting brownie)
Holden Arboretum
Kirtland, Ohio
28 June, 1941

HYPHOLOMA
No 5790
Hypholoma aggregatum Pk
North Chagrin Reservation
Willoughby Hills, Gates Mills, Mayfield, Ohio
4 September, 1937

Walters noted: "North Chagrin Met Park. A form with scales scarcely evident. According to Mr. Beardslee, this is the form which Peck called H aggregatum."

LACTARIUS
No 5822
Lactarius affinis Pk
(Kindred milk cap)
Station touristique Duchesnay
Sainte-Catherine-de-la-Jacques-Cartier, Québec
26 August, 1938

The *Lactarius affinis* is found throughout northeastern North America, identifiable by its wide, sticky, pale ochre-yellow or pink cap and its distinctive gills and thick stalk. The edibility of the kindred milk cap is debated, but it has a highly acrid taste that certainly makes it unpalatable.

LACTARIUS
No 5825
Lactarius deceptivus Pk
(Deceiving milk cap)
Station touristique Duchesnay
Sainte-Catherine-de-la-Jacques-Cartier, Québec
24 August, 1938

The specific epithet *deceptivus* in this mushroom's name, translated from the Latin to "deceptive," possibly alludes to this fungus's contrasting appearance of young and old fruit bodies.

It was first described by American mycologist Charles Horton Peck in 1885. Peck said of the mushrooms' edibility that "An experiment of its edible quality was made without any evil consequences," though mycologists have since deemed it inedible largely due to its strongly acrid taste.

239

LACTARIUS
No 5823
Lactarius chrysorrheus Fr
(Yellowdrop milk cap)
North Chagrin Reservation
Willoughby Hills, Gates Mills, Mayfield, Ohio
4 November, 1941

The *Lactarius chrysorrheus*, commonly known as the yellowdrop milkcap,
is pale salmon with darker markings in rings or bands across the cap.
It is widely considered a poisonous mushroom, containing toxins that can
result in potentially severe gastrointestinal symptoms.

240

LACTARIUS
No 5832
Lactarius insulsus Fr
North Chagrin Reservation
Willoughby Hills, Gates Mills, Mayfield, Ohio
29 July, 1947

LACTARIUS
No 6364
Lactarius deceptivus Pk
(Deceiving milkcap)
North Chagrin Reservation
Willoughby Hills, Gates Mills, Mayfield, Ohio
1 August, 1951

242

LACTARIUS
No 5828
Lactarius helvus Fr
(Fenugreek milkcap)
Sainte-Anne-de-Bellevue, Québec, Canada
26 August, 1941

Found across coniferous woodland in Europe and occasionally North America, the fenugreek milkcap is named after its distinctive smell, which has also been likened to licorice, celery, or Maggi instant soups. The North American variety has a sweeter smell, compared to maple syrup, and in Québec it is known as the maple milky cap, *lactaire à odeur d'érable*.

LEPIOTA
No 5865
Lepiota felina Fr (Left), *Lepiota acerina Pk* (Right)
(Cat dapperling mushroom)
Willoughby Hills, Gates Mills, Mayfield, Ohio
19 October, 1941

The *Lepiota felina* or cat dapperling mushroom is common across
Britain and Ireland, mainland Europe, and North America. Its common
name is likely a reference to the unique leopard-like spots on its cap.

LEPIOTA
No 5870
Lepiota clypeolaria Fr
Thompson, Ohio
29 September, 1938

LEPIOTA
No 5871
Lepiota cristata Fr
(Stinking dapperling, stinking parasol)
Holden Arboretum
Kirtland, Ohio
26 September, 1938

Lepiota cristata's common names include the stinking dapperling and stinking parasol, a result of its distinctive unpleasant odor which has been described variously as fishy, rubbery, and fungus-y. In spite of this, the stinking dapperling has been described as having a mild taste, though its toxicity is debated, as individuals have reported gastrointestinal symptoms following consumption.

246

LEPIOTA
No 5873
Lepiota friesii Lasch
(Freckled dapperling)
North Chagrin Reservation
Willoughby Hills, Gates Mills, Mayfield, Ohio
2 September, 1937

Walters noted of this specimen: "North Chagrin Met Park. Differs from *L. acutaesquamosa* chiefly in the abundantly forked gills."

The freckled dapperling mushroom, *Lepiota friesii*, has previously been described by some as edible; however, this mushroom has also been shown to cause alcohol intolerance and is potentially poisonous.

LEPIOTA
No 5881
Lepiota procera Fr
(Parasol mushroom)
North Chagrin Reservation
Willoughby Hills, Gates Mills, Mayfield, Ohio
17 September, 1938

248

LEPIOTA
No 5882
*Lepiota rhacodes (Vitt) Quel, Lepiota brunnea
(Farl & Burt)*
(Shaggy/brown parasol)
Location unknown
6 September, 1942

Walters noted: "The name brunnea is probably preferable, if, as Dr. Overholts says, it is to be maintained, since it is apparently the American form of rhacodes."

LEPTONIA
No 5888
Leptonia lampropoda Fr
Holden Arboretum
Gates Mills, Ohio
4 July, 1940

250

LEPTONIA
No 5887
Leptonia asprella Fr
North Chagrin Reservation
Gates Mills, Ohio
23 June, 1939

LEPTONIA
No 5886
Leptonia asprella Fr
Thompson, Ohio
29 September, 1938

252

MARASMIUS
No 5901
Marasmius erythropus Fr
North Chagrin Reservation
Willoughby Hills, Gates Mills, Mayfield, Ohio
1 August, 1938

MARASMIUS
No 5898
Marasmius cohoerens Fr Bres
(Bristled parachute)
North Chagrin Reservation
Willoughby Hills, Gates Mills, Mayfield, Ohio
29 August, 1942

254

LACTARIUS
No 5854
Lactarius volemus Fr
(Weeping milk cap)
North Chagrin Reservation
Willoughby Hills, Gates Mills, Mayfield, Ohio
22 July, 1945

LEPIOTA
No 5903
Marasmius glabellus Pk
North Chagrin Reservation
Willoughby Hills, Gates Mills, Mayfield, Ohio
1 August, 1938

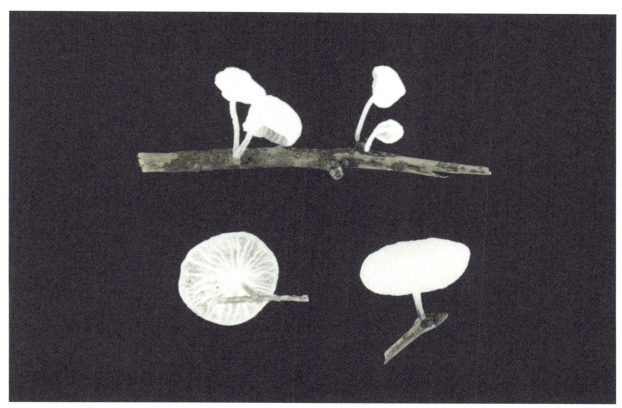

(Top) LEPIOTA, No 5904. *Marasmius magnisprous murrill.* Great Smoky Mountains National Park (Agency: U.S.), Gatlinburg, Tennessee. 17 August, 1939. (Bottom) LEPIOTA, No 5907. *Marasmius oreades Fr.* Location unknown. 16 July, 1940.

(Top) MISC, No 5706. *Exidia gelatinosa (Bull.) Schroct.* on *Ostrya virginia* (witches butter fungus). North Chagrin Reservation, Willoughby Hills, Gates Mills, Mayfield, Ohio. 17 May, 1942. Walters noted: "The corrected name, according to Markin, is *Exidia recisa* (S.F. Gray) Fries." *(Bottom)* BOLETUS, No 5497. *Boletus subtomentosus.* L. North Chagrin Reservation, Gates Mills, Ohio. Date unknown.

MISC
No 5707
Favolus canadensis (Klotzsch)
North Chagrin Reservation
Willoughby Hills, Gates Mills, Mayfield, Ohio
26 May, 1939

BOLETUS
No 5467
Boletus felleus (Bull)
(Bitter bolete)
North Chagrin Reservation
Willoughby Hills, Gates Mills, Mayfield, Ohio
30 June, 1935

Walters nicknamed this specimen "the Bitter Boletus (probably forma plumbeoviolaceus snell)."

CLITOCYBE
No 6387
Clitocybe clavipes Fr
(Club foot)
Holden Arboretum
Kirtland, Ohio
2 November, 1941

MORELS
No 5934
Morchella conica
(Black morel)
Location unknown
Date unknown

With approximately 70 species documented worldwide, fungi within the morchella family, or "true morels," are a genus of ascomycetes (sac fungi) that are highly prized in many European cuisines, particularly French, Basque, and German. With a spring fruiting season—they can be found in both fields and woodlands, and particularly favor ashy soil—true morels have, historically, been so coveted that in the 18th century, heath fires had to be restricted in Germany and Provence, as peasants were reportedly setting the moors on fire to encourage morels to grow. Notable for their distinct ridged and pitted cap shape, often likened to that of a honeycomb, edible morels (which must be cooked properly—they are never to be eaten raw or undercooked) are easily confused with species of "false morels," such as *Gyromitra esculenta* and *Verpa bohemica,* both of which can induce serious, even fatal, bodily reactions.

(Top) MORELS, No 5933. *Morchella conica* (black morel). Location unknown. Date unknown. *(Bottom left)* MORELS, No 5935. *Morchella crassipes* (bigfoot morel). Location unknown. Date unknown. *(Bottom right)* MORELS, No 5937. *Morchella deliciosa* (white morel). Location unknown. Date unknown.

MORELS
No 5938
Morchella esculenta
(Yellow morel)
Location unknown
Date unknown

Similar to the black morel, the *Morchella esculenta* is a popular edible mushroom, but must be cooked before eating, as raw morels have a gastrointestinal irritant that is removed by parboiling or blanching. They are typically served fried in butter or stuffed with meats and vegetables before baking.

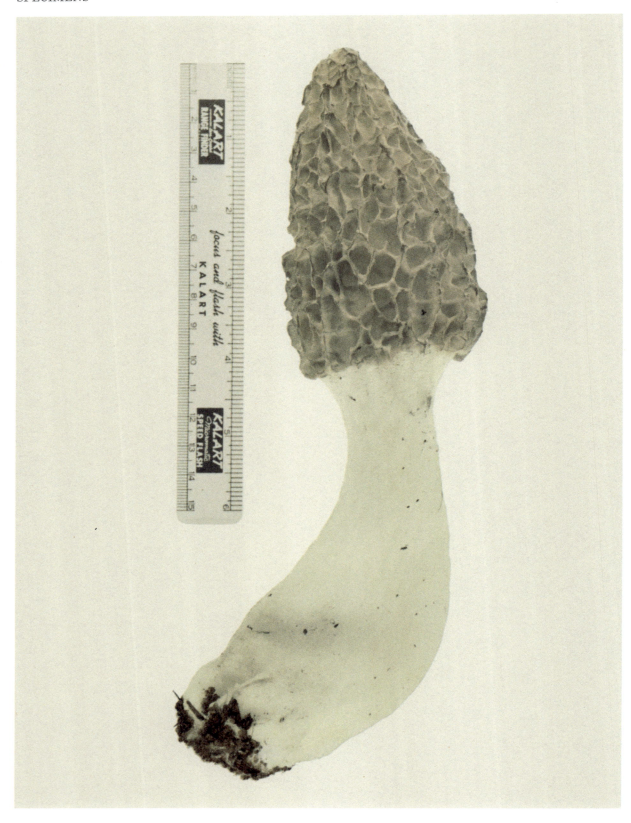

MORELS
No 5936
Morchella crassipes
(Bigfoot morel)
Location unknown
Date unknown

Walters occasionally included rulers in his studies when relevant, as illustrated on this and the next page.

266

POLYPORUS
No 6419
Polyporus radicatus (Schw)
North Chagrin Reservation
Willoughby Hills, Gates Mills, Mayfield, Ohio
1 October, 1947

HYPHOLOMA
No 5800
Hypholoma rugocephalum atk
North Chagrin Reservation
Willoughby Hills, Gates Mills, Mayfield, Ohio
3 September, 1940

268

CORTINARIUS
No 5633
Cortinarius corrugatus Pk
North Chagrin Reservation
Willoughby Hills, Gates Mills, Mayfield, Ohio
17 June, 1941

MISC
No 6052
Pholiota mutabilis Fr
North Chagrin Reservation
Willoughby Hills, Gates Mills, Mayfield, Ohio
17 May, 1937

Walters noted "There is some doubt as to this identification. Caps average 2.8cm. Gills minutely crenulate in some cases scarcely reaching margin. Stems 8cm × 5mm. The cops are slightly viscid, distinctly hygrophanous pellucid-striate when moist and have a thin, separable pellicle. Spores 8 × 4.5m, inequilateral. Gill edges thick with long, narrow, twisted sterile cells. Sometimes slightly enlarged at apex, 4m wide. Later specimens showed a much more orange-buff tinge."

MISC
No 6055
Pholiota praecox Fr
North Chagrin Reservation
Willoughby Hills, Gates Mills, Mayfield, Ohio
3 June, 1937

272

MISC
No 6040
Pholiota caperata Fr
Holden Arboretum
Kirtland, Ohio
10 November, 1936

MISC
No 6038
Pholiota adiposa Fr
North Chagrin Reservation
Willoughby Hills, Gates Mills, Mayfield, Ohio
19 October, 1937

MISC
No 6037
Pholiota acericola Pk
North Chagrin Reservation
Willoughby Hills, Gates Mills, Mayfield, Ohio
27 July, 1936

274

MISC
No 6371
Pluteus coccineus Massee
North Chagrin Reservation
Willoughby Hills, Gates Mills, Mayfield, Ohio
21 August, 1950

ENTOLOMA
No 5699
Entoloma lividum Fr
Holden Arboretum
Kirtland, Ohio
8 September, 1940

Walters noted that this specimen had cream-yellow gills.

276

PSATHYRELLA
No 6155
Psathyrella microsperma Pk
Acacia Reservation
Lyndhurst, Ohio
2 November, 1941

RUSSULA
No 6167
Russula mariae Pk
North Chagrin Reservation
Gates Mills, Ohio
23 June, 1939

The *Russula mariae* is recognizable by its pink or violet cap, though its color can vary to include yellow. It is an edible mushroom, though its taste is often acrid.

278

RUSSULA
No 6171
Russula virescens Fr
(Green-cracking russula, quilted green russula,
green brittlegill)
North Chagrin Reservation, Gates Mills, Ohio
29 June, 1939

The *Russula virescens* can be recognized by its pale green cap. Its name
is derived from the Latin *virescens,* meaning "becoming green." The fungi's
characteristic cracked pattern has earned it the common names green-
cracking russula and quilted green russula.

Russula virescens is considered one of the most edible mushrooms of the
Russula genus. Used in traditional herbal medicines in Asia, they are typically
sautéed, fried, grilled, or eaten raw in salads across Europe and particularly
in Spain.

CORTINARIUS
No 5629
Cortinarius
Holden Arboretum
Kirtland, Ohio
7 September, 1941

Walters noted: "Holden Arboretum. Unknown. Probably new species. Cap 3.4cm, margin inrolled, color a dark lavendar [*sic*], covered w/ appressed scales, but somewhat shining. Gills a rich orange buff, rather broad, with a distinct long decurrent tooth, close, thin. Odor & taste not noticeable."

MISC
No 6363
Ithyphallus ravenelii
North Chagrin Reservation
Willoughby Hills, Gates Mills, Mayfield, Ohio
20 September, 1949

DICTYOPHORA
No 5677
Dictyophora duplicata
(Oak mazegill)
North Chagrin Reservation
Willoughby Hills, Gates Mills, Mayfield, Ohio
Date unknown

282

Named for their combination of pungent odor and phallic form, fungi within the stinkhorn (Phallaceae) family are most commonly found in coniferous and broadleaf woodlands. Their foul smell, typically that of carrion or dung, emanates from the spore mass, or gleba (the fleshy, spore-bearing inner mass of certain fungi) that sits at the end of their stalk, known as the receptaculum. This odor attracts insects such as flies or beetles, which help to disperse the fungi's spores. First developing as primordia, or "eggs" (spherical structures that may be fully or partially buried underground), stinkhorns lie dormant until their fruiting season, which runs from summer to late fall. Species within the Phallales order, such as *Phallus indusiatus* (also known as the basket stinkhorn or veiled lady), are considered some of the world's fastest-growing organisms, with growth measurements recorded at 0.5 centimeters per minute, often accompanied by what has been described as a "crackling" sound due to the speed of their growth. This specific kind of morphogenesis has recently been termed a form of "cellular origami."

STINKHORN
No 5941
Mutinus curtisii (Berk)
North Chagrin Reservation
Willoughby Hills, Gates Mills, Mayfield, Ohio
11 August, 1935

STINKHORN
No 5942
Mutinus curtisii
North Chagrin Reservation
Willoughby Hills, Gates Mills, Mayfield, Ohio
Date unknown

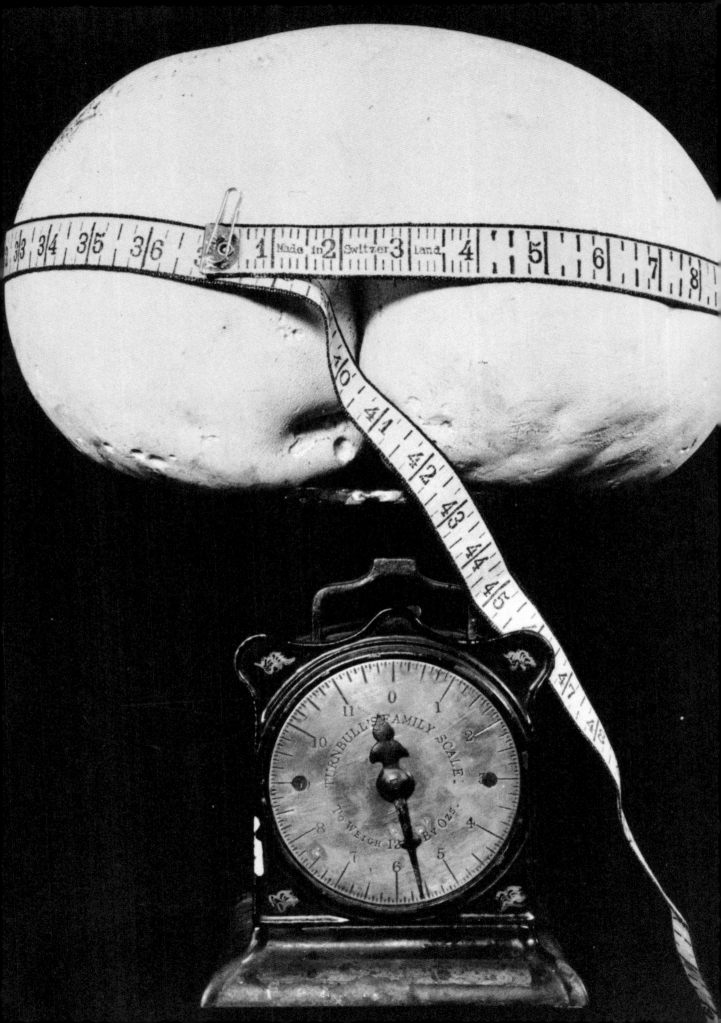

The Giant Puffball, the Amateur, and His Love of Dogs

by Siôn Parkinson

There's something wonderfully silly about coming across a giant puffball in the woods or patch of grassland. For one, their enormous size. The giant puffball, as its name suggests, is one of the largest fungal fruiting bodies (or mushrooms, as they are more commonly known), and at its most mature stage typically balloons anywhere between 8 to 20 inches in width. A mature giant puffball can produce up to 7 trillion spores. To try to put this obscene output into some sort of human perspective, approximately 117 to 125 billion people have lived on Earth in the whole of human history.[1] Or to put this another way, 7 trillion is nearly 60 times the current global population. Which begs the question: Why is the forest floor not carpeted with these marsh-mallowy monsters? The fact is, less than 0.001 percent, and maybe as little as 0.00001 percent, of the giant puffball's spores germinate, meaning the excesses of this mushroom, like those of most other fungi, are another classic example of nature's strategy to ensure survival by overwhelming the world.

Another reason the giant puffball seems so absurd is its sudden and striking appearance, looking, as it does, like a brain, buttocks, or bloated belly protruding from the earth. Their resemblance to plaster casts of human body parts reminds us of anatomical votive offerings deposited in the woods under the moonlight—bones, limbs, and organs excised to beseech some forgotten god of antiquity to fulfill an erotic fantasy, say, or to cover a bald patch, cure a stomachache, or answer a prayer for fertility. Their boisterous calls for attention seldom go unnoticed by those willing to wander into the woods—such as the mushroom hunter.

A series of three photographs by nature enthusiast Maurice Burke Walters crouching next to a bunch of giant puffballs in North Chagrin, Cleveland Metropolitan Park, conveys this sense of care and concern for the overlooked or out-of-the-ordinary aspects of the natural world (fig. 2). The photos were taken on October 22, 1941, precisely 50 days before Germany

287

(*Previous spread*) Fig. 1: *Calvatia gigantea*, 22 October, 1941. Walters's giant puffball on a Turnbull's Family scale, with tape measure and paper clip. (*Above*) Fig. 2: *Calvatia gigantea*, 22 October, 1941. A presumably self-timed photograph. Walters crouches next to a bloom of giant puffballs in a woodland clearing. The specimen in the foreground appears to be the same mushroom he photographed in his studio.

288

and Italy declared war on the United States, an event that prompted the US to reciprocate and fully commit to the Allied war effort on both the European and Pacific fronts. Cleveland's steel mills were reaching near capacity in order to supply materials for tanks, ships, and weapons. At the same time, the Cleveland Metroparks system, established in 1917, provided a respite for the city's inhabitants from the turmoil of war enveloping the globe. Reservations like North Chagrin, which includes Ohio's only old-growth forest, are where Walters spent much of his time during the 1940s and '50s. He led nature walks for young people and other community groups here, especially bird-watchers, as well as venturing out alone with his camera. This is where we find him squatting beside a clutch of outrageously large puffballs in a woodland clearing.

Dressed in a white shirt and tie beneath a V-neck sweater, with wool trousers, leather boots, and a jacket, the 50-year-old Walters—who is slightly balding—peers at the ground through round, steel-rimmed spectacles. His faint smile reflects a mix of curiosity and amusement as he delicately examines two of the larger specimens. Their odd shapes, warped by their rapid growth around the stems of the sycamore saplings sprouting up everywhere, resemble blobs of redaction fluid. Solid and chalky-white against the leaf litter, they evoke a heavily censored text, with each blank form obscuring an expletive.

By inserting himself into the (presumably self-timed) photograph, Walters not only demonstrates the scale of his remarkable find but also his sense of delight—an emotion that for most mushroom enthusiasts is often experienced privately. His demeanor invites the viewer to share in his wonder at the fungal world. This passion is evident across Walters's captivating and occasionally peculiar mushroom photographs, which are preserved in the United States Department of Agriculture (USDA) National Fungus Collection.

Sometimes the scientific effort to describe how things truly are inadvertently crosses into art. In an attempt to study his puffballs more closely, Walters removes one especially rude-looking specimen to his home studio, where he photographs it against a black backdrop on a Turnbull's Family upright scale (fig. 1). The mushroom is presented wrapped in a dressmaker's tape measure, fastened at the front with a paper clip. The configuration of oversize fungus with familiar household implements allows Walters to communicate multiple characteristics of his subject simultaneously: the mushroom's relative scale, its mass (5 pounds 9 ounces), and its circumference (37 inches). These measurements make Walters's giant puffball roughly equivalent in size and weight to the skull of an Asian elephant. And while this comparison may still seem pretty esoteric, its purpose is to convey the mammoth proportions of Walters's subject, along with the uncanny effect that arises when natural history objects are displayed outside of their original context. In this sense, Walters's picture evokes the kinds of images one might encounter in the dimly lit corners of a natural history or medical museum, where labels or captions to the exhibits have long since fallen off.

Of all his mycological photographs, Walters's studio portrait of the giant puffball stands out for being seriously strange. The image demands a second or even third glance to puzzle out exactly what is happening. It is suggestive, in the sense that it evokes far more than what it explicitly shows— much like the mushroom itself is but a fleeting visible emergence of the vast, hidden mycelial network beneath the ground. Though the main purpose of the image is to show the mushroom's scale, weight, and dimensions, Walters's puffball portrait is bizarre because it communicates more than what the photographer intends to represent.

The visual effect of combining ordinary domestic or manufactured objects with the *foreign*—literally "out of doors"—is uncanny, bordering on the absurd. In this way, Walters's composition of a massive mushroom on a kitchen scale with tape measure and paper clip recalls the Dadaist "readymades" of the 1920s and '30s, where everyday objects were presented, with little alteration, as art. For example, Man Ray's *Indestructible Object,* originally created in 1923 and modified through 1950, which, with its mechanical metronome and cutout eyeball (belonging to photographer Lee Miller, Ray's former partner) attached with a paper clip to its inverted pendulum, bears a striking resemblance to Walters's setup. Like Man Ray's "object to be destroyed" (the artwork's original title), the giant puffball's bulbous shape, outlandish proportions, and reproductive potential—plus the contradiction in tone between the "serious" matter of scientific observation and measurement and the functionless-ness of art—results in an image that is, quite frankly, bizarre.

In the binomial system introduced by Swedish scientist Carl Linnaeus, the scientific name of the giant puffball, *Calvatia gigantea,* translates roughly as "gigantic bald one." This humanizing, almost affectionate epithet reflects how mycologists—specialists in fungi—often name and describe the natural world in relation to themselves or their immediate surroundings.

When we come across something freakishly large in nature that we want to share with others, we tend to use whatever, or whoever, is at hand to help us show scale. Search the internet for images of giant fungi, and you will find countless examples of pictures where mushroom hunters have placed everyday objects next to or on top of their mushrooms. Take, for example, Walters's photo of an oversize bitter bolete with his pocket watch and pipe placed on top (fig. 3). The cap of this species of mushroom typically grows up to 6 inches in diameter. Walters's is a whopping 10½ inches. As well as showing scale, these objects help to situate the photograph in a particular time and place. But when nothing else is available, we often turn to our own hands to serve the purpose. Hands pointing, patting, stroking, squashing, cupping, or girdling puffballs on the ground or in the air, as if forming quotation marks around the answer to one inevitable question: "How big?!"

Alongside his own hands and various odds and ends emptied from his pockets, Walters occasionally includes people—and sometimes even his pet dog—in his pictures of larger mushrooms. At times it seems as though Walters selects his sitters, human or otherwise, because of their physical resemblance to the shape of the mushroom itself. One photo features someone identified as Bob Welchans with an "extra large specimen" of an elfin saddle (fig. 4). Welchans was a film director for Cinecraft Productions, a Cleveland-based company specializing in industrial films that is still active today. He was also a bird enthusiast and, like Walters, a member of the Cleveland Bird Club, now the Audubon Society of Greater Cleveland. In the image, Welchans's silver bouffant and wavy quiff bear an uncanny resemblance to the wrinkled gray cap and fluted stem of the elfin saddle. Walters captures the specimen in another photograph taken on the same day and on the same kitchen scale that features

290

(Top) Fig. 3: "Boletus felleus." An enormous bitter bolete (current name *Tylopilus felleus*) with pocket watch and pipe. Date unknown. *(Bottom left)* Fig. 4: "Bob Welchans with our extra large specimen of *Elvela underwoodi* [*sic*] Seaver, collected at North Chagrin Met. Park in the 'big woods' near 'upper 40.'" 21 May, 1944. *(Bottom right)* Fig. 5: Walters's and Welchans's oversize elfin saddle fungus on kitchen scale with ruler. Date unknown.

(Top) Fig. 6: "Polyporus berkeleyi & Ginny." Berkeley's polypore fungus (current name *Bondarzewia berkeleyi*) with Walters's pet dog Ginny. 27 August, 1946. *(Bottom)* Fig. 7: The last picture Walters took of his dog Ginny before the family had to put her down. 20 April, 1951.

292

in his picture of the giant puffball, this time replacing the tape measure with a ruler (fig. 5).

Another photograph features Herbie Rickert, a teenage member of the Cleveland Bird Club, perched on a wooden bench or fence with a large oyster mushroom, known as the hairy panus, balanced in his lap (fig. 8). Rickert, with his pasty complexion, messy crew cut, and mottled leather jacket, subtly mirrors the awkward, uneven folds of the pale mushroom.

In perhaps the most striking of these portraits, Walters includes his German shepherd, Ginny, posing with an enormous fan-shaped polypore—a parasitic fungus that causes a condition known as "butt rot" in the base of trees (fig. 6). Positioned side by side on a freshly mown lawn, the monumental form of the polypore dwarfs Ginny, its shadow engulfing almost her entire body, save for her golden ruff and forepaws, which catch the summer sunlight. With her head cocked to one side, Ginny's presence in the photograph expresses something beyond mere scale. Curiosity. Affection, maybe. Perhaps even love.

Ginny reappears seven years later in a photograph of a woodland scene taken by Walters in April 1950 (fig. 7). A circle of bare-branched trees is reflected in a vernal pool, their stark forms mirrored in the still water. About 20 or 30 feet away, a fallen tree trunk lies on its side, presumably brought down by a spring storm. It is difficult to make out, but perched partway along the trunk near the tree's exposed root plate is a seated figure wearing a dark hat, gazing across the pool directly at the camera. This figure is likely Walters's wife, Gertrude Chamberlain Walters, identified on the reverse of the photograph simply as "GCW." She is stroking the chest of a dog—Ginny—who is standing atop the trunk beside her, and who is also staring at the camera. It is a peaceful scene, composed in the literal sense of being deliberately set up. You can almost hear Walters's voice echoing through the woods, calling across the pond to catch the attention of his wife and dog, just before the click of the shutter.

A handwritten note on the back of the photograph offers a clue as to why this image might have been so carefully staged (fig. 6). The note reads: "Ginny's last picture, taken one week before she was put away. At 'Salamander Pool' (just beyond the 'Fairy Shrimp Pool')."

The image appears out of place—amateur, even—because unlike the rest of Maurice Walters's photographs in the US Department of Agriculture's collection, this one is not really about mushrooms at all. The deeply personal subject matter feels doubly out of place in what is ostensibly a collection devoted to specialist descriptions of fungi. This is amateur photography in the truest sense of the word. After all, this is why the camera was adopted by so many: not to document nature's curiosities or produce risqué pictures of oddly shaped things, but to capture and commemorate what we cherish most, be it mushrooms or mutts.

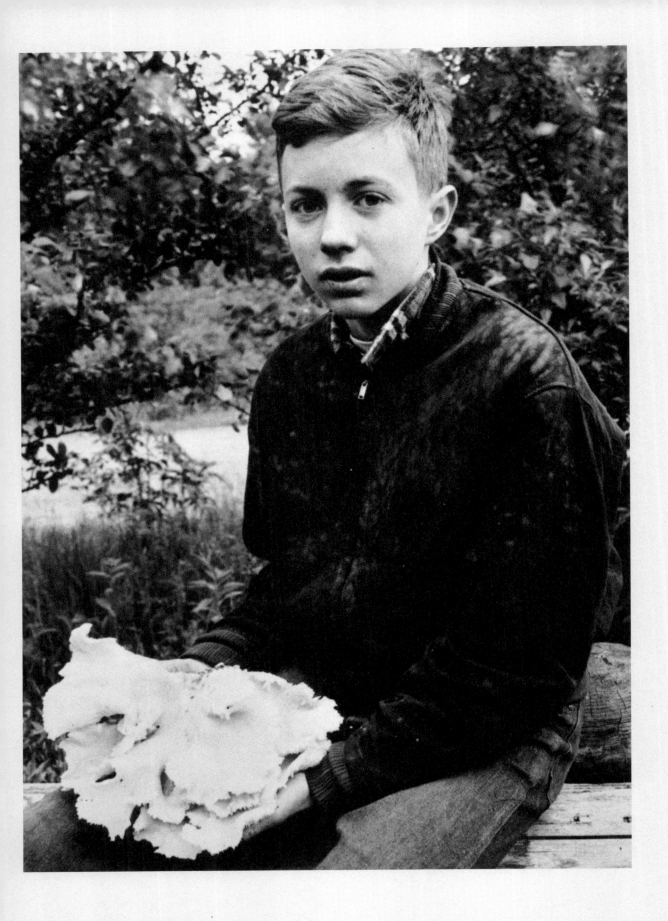

Fig. 8: Herby Rickert with a hairy panus, or hairy oyster mushroom (current name *Lentinus strigosus*). 24 May, 1942. In October 1950, eight years after this photo was taken, Rickert and Walters both contributed to the *Cleveland Bird Calendar,* a journal focused on documenting seasonal bird observations in the Cleveland area and northeastern Ohio.

Spore Time

The Fungal Photographic Experiments of Maurice B. Walters

by Siôn Parkinson

Simple instructions for making a spore print: Take a mature mushroom, typically one with gills, but any will do—even one purchased from your local grocery store. Snip off its stalk, and place the cap gill-side down against the surface of a piece of black or dark gray paper. (A useful tip here is to use paper half of which is printed black and the other half left white, positioning the mushroom cap directly over the dividing line. This setup ensures that spores of different colors or shades, which may not always show up on solid white or black paper alone, will be more visible.) Cover the cap with an upturned bowl; choose a bowl that's both large enough to fully enclose the cap and opaque enough to block out light. This step simulates the nighttime conditions under which the mushroom would naturally release its spores. Allow it to sit undisturbed for a few hours or overnight. Carefully remove the bowl, then the cap (which you may dispose of), to reveal the startling results (fig. 1). *Et voilà!*

Spore prints are an essential tool for distinguishing between mushroom species, as they can reveal critical differences in spore color, a key identifying feature of mushrooms. For example, when differentiating between older specimens of the delicious violet-colored wood blewit (*Collybia nuda*) and some species of purple webcaps in the genus *Cortinarius,* many of which are poisonous, spore prints are particularly helpful. (Webcaps typically produce rusty-brown-colored spores, while blewits tend to produce a pale-lilac spore print.) But beyond individual motives for avoiding an upset stomach, damaging your liver, or dodging death, spore prints can be enjoyed simply for their striking visual effect, which, like the steamy halo left by a warm hand on a highly polished piano lid, leaves a trace of a thing no longer there.

While photography creates images by concentrating and directing light onto a surface for a fraction of a second, spore prints are positive impressions of mushrooms, formed not by light but through the gradual, natural scattering of microscopic spores onto paper over several hours. Fungal spores are biologically different from plant seeds in both structure and function. Spores

297

(Previous spread) Fig. 1: "Spore print and sporophore. Collybia radicata." Also known as the deep root mushroom (current name *Hymenopellis radicata*). 28 September, 1950. *(Top)* Fig. 5: "Coprinus." Side-by-side photographic prints depicting a cluster of inkcaps (species not identified). 16 September, 1949. *(Bottom)* Figs. 4 & 5: Netted stinkhorn (current name *Phallus duplicatus*).

298

are simple, minuscule, usually single-celled units that can grow into a new fungus, often without the need for sexual fusion between different fungal cells. The word spore comes from the ancient Greek *spora,* meaning "seed," which relates to the act of and season for planting seeds by sowing or scattering loosely on or in the earth. Spores—and by extension spore prints—thus represent a kind of "seed time"; that is, they embody the act of reproducing the primary object (the mushroom) through the slow dispersal of microscopic particles onto a surface.

Adjusting this slightly, what might *spore time* signify in the context of creating images of or with mushrooms? What does this process reveal about fungal temporality and reproduction? And how does spore time intersect with photography's process of doubling, in which photographic printing transforms a single image into multiples, enabling potentially infinite iterations of an original?

These are the kinds of questions Maurice B. Walters grappled with—practically, if not theoretically—while experimenting with various photographic techniques to capture and convey mushroom morphology. Beyond spore prints, a closer examination of Walters's black-and-white and hand-colored prints reveals the diverse methods he repeatedly tested and refined to effectively depict the distinctive characteristics of each species (figs. 2–5). For instance, his monochrome print of an upside-down polypore (which, one can just make out, is suspended by a piece of string tied to the base of the mushroom's stalk where it splits to form twin fruiting bodies) was later cut out, inverted, and tinted, transforming the composition to highlight the pigmented variation in its fleshy, fanlike fronds (figs. 6–7). Similarly, his color tests with the bloodred beefsteak mushroom demonstrate his meticulous approach: two hand-tinted prints, placed side by side—one on a black background and the other on white—highlight his painstaking efforts to capture the exact tonal subtleties of this meaty species (figs. 8–9). These examples showcase not only Walters's scientific rigor but also his sensitivity to the aesthetic potential of his medium.

Elsewhere, Walters explores the illusory possibilities of photographic collage, combining front and rear views, or cap and underside, of the same specimen into a single composition. His images of polypores (figs. 10–11), scalycaps (fig. 12), snow morels (fig. 13), and others resemble topsy-turvy magic tricks, where more than one side of the mushroom appears visible at the same time. As such, and in Walters hands, spore time does not only affect temporality—the way we perceive a mushroom's existence within time—but also spatiality; that is, the way we perceive, experience, and understand the mushroom in its environment. Spore time, in this sense, alters our spatial awareness, asking us to engage with mushrooms not as isolated, fixed objects but as dynamic, multidimensional organisms that demand a more complex, shifting view. In other words, spore time demands that we slow down our gaze, and that we take a 360-degree view of the natural world so that we may see it from all sides.

This is nothing new. The history of mycology is closely intertwined with the history of printmaking. From the first woodcut of a bunch of nondescript "agarics" in the late 1400s to the copperplate engravings in the appendix Fungorum, dedicated to the fungi of Central Europe, in Carolus Clusius's *Rariorum plantarum historia* (1601), the evolution of print techniques have been central to visualizing mushrooms.[1] A hundred years later, the introduction of color to printing added new dimensions, and the rise of lithographic prints in the 1800s allowed for more detailed and vivid depictions.

By the mid-20th century, photography had replaced lithography as the dominant medium for capturing the finer details of fungi. Prior to this shift, monochrome (black-and-white) photography remained the primary method, due in part to the high cost and limited availability of color negative processing. As a result, amateur photographers often resorted to coloring their prints by hand, applying watercolors or other dyes with brushes, a cost-effective technique that added tonal variation and subtle color gradients. Today, the tradition of scientific illustration continues, with cutting-edge tools ranging from 3D microscopy to digital scanning with smartphone cameras. These modern techniques offer unprecedented levels of color, magnification, and detail, making it possible for almost anyone to capture fungi—whether in the lab or in the field—with remarkable clarity. Such advancements in imaging technology mirror the historical evolution of mushroom visualization, continually expanding our capacity to zoom in on even the smallest of fungal features.

And yet: Despite the advancements in imaging technologies, there remains something intimate and uniquely attuned to the temporospatial qualities of mushrooms in Walters's hand-colored photographs. His work, rooted in a commitment to empirical observation and veracity, but also experimentation, captures not only the physicality of the specimens but their very essence in time and space. In an era where digital tools dominate, Walters's painstaking methods—his approach to spore time—serve as a reminder of the value of slowing down and immersing oneself in the very tiny details. As such, it is only fitting that his humble spore prints and hand-colored photographs, some of them almost a hundred years old, now receive the recognition they deserve.

(Top) Figs. 4 & 5: "Fistulina hepatica." Two colorations of a beefsteak fungus showing slight differences in hue and shade.1 October, 1940. *(Bottom left)* Fig. 6: "Polyporus betulinus." Combined top and underside view of a birch polypore specimen (current name *Fomitopsis betulina*). 25 August, 1941. *(Bottom right)* Fig. 7: "Polyporus [*illegible*]." Combined top and underside view of a single polypore, possibly the tuberous polypore (*Polyporus tuberaster*). Date unknown.

Figs. 10 & 11: "Polyporus berkeleyi." Berkeley's polypore (current name *Bondarzewia berkeleyi*). 27 July, 1936.

(*Top*) Fig. 12: "Flammula flavida." Front and back view of single specimen of scalycap (current name *Pholiota flavida*.) 27 October, 1946. (*Bottom*) Fig. 13: "Gyromitra gigas." Combined front and back view of single specimen of a snow morel, calf brain, or bull nose fungus. (Its scientific name roughly translates as "giant turban.") Date unknown.

Appendix

Glossary

A

AGARICALES
An order of fungi in the class Agaricomycetes. One of the most diverse orders, it is currently known to contain approximately 30 families, 350 genera, and some 10,000 species.

AMADOU
A spongy material made by drying certain bracket fungi, such as *Fomes fomentarius* (Iceman fungus).

ANASTOMOSIS
The fusion that occurs between branches of hyphae to make a network.

ASCOMYCOTA
A phylum of fungi characterized by their sac-like structure (the ascus) where sexually produced spores are formed.

ASCUS
A sac-like structure where ascospores are produced.

B

BASIDIOMYCOTA
A phylum in the Fungi kingdom that includes jelly and shelf fungi, mushrooms, puffballs, stinkhorns, certain yeasts, rusts, and smuts.

BASIDIUM
The organ in the phylum Basidiomycota that bears basidiospores (sexually reproduced bodies).

BASIONYM
Within binomial nomenclature, the original name on which a new name is based.

BINOMIAL NOMENCLATURE
The system of nomenclature in which two terms are used to denote a species of living organism, the first one indicating the genus and the second the specific epithet. Swedish biologist Carl Linneaus is responsible for the formalization of binomial nomenclature in the 18th century.

BLIGHT
A plant disease typically caused by fungi such as rusts, smuts, and mildews.

C

CAP
Also known as the pileus, the broad upper part of the fruiting body of a mushroom that bears gills or pores.

CIRCUMBOREAL
Throughout the boreal regions (i.e., Canada, China, Finland, Japan, Norway, Russia, Sweden, and the United States).

CIRCUMSCRIPTION
The content of a taxon, i.e., the delimitation of which subordinate taxa are part of that taxon.

CLASS
A principal taxonomic grouping that ranks above order and below phylum or division.

CONIFEROUS FOREST
A forest in which trees produce cones and have leaves that do not fall during winter.

CORTINA
The thin, web-like veil that extends from the edge of a mushroom's cap (pileus) to its stalk (stipe), especially when immature.

COSMOPOLITAN
A species that can be found worldwide.

D

DECIDUOUS FOREST
A temperate forest in which trees lose their leaves at the end of each growing season.

F

FAIRY RING
A naturally occurring ring or arc of mushrooms that grows from a central mycelium.

FAMILY
A principal taxonomic category that ranks above genus and below order.

FILAMENT
A term used to describe the fine, thread-like structures in fungi.

FOLIICOLOUS
The growth habit of certain lichens, algae, fungi, liverworts, and other bryophytes that prefer to grow on the leaves of vascular plants.

FORAGING
The act of searching for wild food resources, typically fruit, nuts, and fungi, from the outdoors.

FORAY
An outing, the goal of which is to find, identify, and record fungi growing in a particular habitat.

FRUITING BODY
The reproductive (i.e., spore-bearing) structure of a fungus, also known as the sporophore.

FUNGUS
Any of the known species in the Fungi kingdom, which includes yeasts, rusts, smuts, mildews, molds, and mushrooms.

G

GENERIC NAME
The first part of a species's scientific name, which represents its genus. (Ex: The generic name of *Tremella fuciformis*, snow fungus, is *Tremella*.)

GALL
An abnormal growth formed in response to the presence of insect larvae, mites, or fungi on plants and trees, especially oaks.

GENUS
A principal taxonomic category that ranks above species and below family. It is denoted by a capitalized Latin name (ex: *Calvatia*, *Amanita*, or *Phallus*).

GILLS
A hymenophore rib under the cap of some species of mushroom.

GLEBA
The fleshy, spore-bearing tissue of a gasteroid fungus, such as a stinkhorn or a puffball.

H

HOST
An animal or plant on or in which a parasite or commensal organism lives.

HYMENIUM
A fungi's spore-bearing layer of tissue, found in the Ascomycota and Basidiomycota phylums.

HYPHA
The thread-like filament of a mycelium.

K

KINGDOM
The highest category in taxonomic classification.

L

LATEX
A milk-like substance secreted by from certain types of fungi when they are cut or damaged.

LITHOGRAPHY
The process of printing from a flat surface that has been treated to repel ink except where it is required for printing.

M

MILDEW
A conspicuous mass of thread-like white hyphae and fruiting structures produced by various fungi.

MOLD
A conspicuous mass of mycelium and fruiting structures produced by various fungi.

MUSHROOM
Also known as the sporophore, this is the umbrella-shaped fruiting body of certain fungi.

MYCELIUM
The mass of branched, tubular filaments (hyphae) that make up the thallus (undifferentiated body) of a typical fungus.

MYCORRHIZA
A type of symbiotic association between plant roots and fungi in which the fungi colonizes the root tissues of a plant, benefiting the health of both species. Mycorrhizae play an important role in plant nutrition, soil biology, and soil chemistry.

MYCOLOGY
The scientific study of fungi.

N

NEMATODE
Any worm of the phylum Nematoda, such as a roundworm or threadworm.

NEMATOPHAGOUS FUNGI
Carnivorous fungi that are able to trap and digest nematodes.

O

ORDER
A principal taxonomic category that ranks below class and above family.

P

PARASITE
An organism that lives in or on an organism of another species and benefits by deriving nutrients at the host's expense.

PHYLUM
The principal taxonomic category that ranks above class and below kingdom.

PILEUS
The cap of a mushroom.

PLANT PATHOGEN
Disease-causing viruses, fungi, and bacteria that attack plants.

POLYPHYLETIC
An assemblage that includes organisms with mixed evolutionary origin but does not include their most recent common ancestor.

POLYPORE
A bracket fungus in which spores are expelled through the fine pores on its underside.

PORES
The small, sponge-like holes found on the underside of a fungus's cap, which produce spores.

R

ROT
Any of several plant diseases caused by species of soil-borne bacteria, fungi, and fungus-like organisms in the phylum Oomycota.

RUST
A plant disease caused by over 7,000 species of fungi of the phylum Basidiomycota.

S

SAPROTROPHIC
Relating to the process of using chemoheterotrophic extracellular digestion to process decayed—dead or waste—organic matter.

SCLEROTIUM
A compact mass of hardened fungal mycelium that contains food reserves.

SMUT
A plant disease caused by several species of fungi that primarily affects grasses, including corn, wheat, sugarcane, and sorghum.

SPOROPHORE
The spore-bearing structure of a fungus.

SPORES
Microscopic biological particles that allow fungi to reproduce.

SPORE GERMINATION
The process by which a dormant spore (metabolically inactive) transforms into a vegetative state, making it capable of both reproducing and growing.

SPECIES
A group of living organisms consisting of similar individuals capable of exchanging genes or interbreeding.

SPECIFIC EPITHET
The second element in the Latin binomial name of a species, which follows the generic name and distinguishes the species from others in the same genus. (Ex: The specific epithet of *Calvatia gigantea*, the giant puffall, is *gigantea*.)

STALK
The part of a mushroom's fruiting body that elevates its spore-producing layer above the ground.

STIPE
The stem of a fungus.

STROMA (PL. STROMATA)
A mass of fungal tissue that has spore-bearing structures either embedded in it or on its surface.

T

TAXON
A group of one or more populations of an organism seen by taxonomists to form a unit, such as a species, family, or class.

TYPE SPECIES
The species name with which the name of a genus or subgenus is considered to be permanently taxonomically associated.

TAXONOMY
The branch of science concerned with classification, especially of organisms.

TELIUM (PL. TELIA)
A structure (typically yellow or orange) produced by rust fungi during its reproductive cycle. It is exclusively a mechanism for the release of teliospores, which are released by wind or water to infect the alternate host in the rust lifecycle.

TOADSTOOL
The spore-bearing fruiting body of a fungus, typically in the form of a rounded cap on a stalk, especially one that is believed to be inedible or poisonous.

V

VEIL
A membrane attached to the immature fruiting body of a fungus that ruptures in the course of development, either enclosing the whole fruiting body (universal veil) or joining the edges of the cap to the stalk (partial veil).

W

WEBCAPS
A colloquial name for fungi within the genus *Cortinarious*, all of which possess a cortina (veil) that connects the pileus (cap) to the stipe (stem) when they are immature.

Y

YEAST
Any of approximately 1,500 species of single-celled fungi, most of which are in the phylum Ascomycota (only a few belong to the phylum Basidiomycota).

Index

Further Reading

Alphen, Ernst van. (2014) *Staging the Archive: Art and Photography in the Age of New Media.* Reaktion Books, London.

Arora, David. (1991) *All That the Rain Promises and More...* Ten Speed Press, Berkeley, CA.

Davidson, Alan. (1999) *The Oxford Companion to Food.* Oxford University Press, Oxford, UK.

Findlay, W.P.K. (1982) *Fungi: Folklore, Fiction, & Fact.* The Richmond Publishing Company, Richmond, UK.

Fine, Gary Alan. (1998) *Morel Tales: The Culture of Mushrooming.* Harvard University Press, Cambridge, MA.

Leith Gollner, Adam, et al. (2021) An *Illustrated Catalog of American Fruits & Nuts: The US Department of Agriculture Pomological Watercolor Collection.* Atelier Éditions, Los Angeles.

McKnight, Vera B. and Kent H. McKnight. (1987) *A Field Guide to Mushrooms: North America.* Houghton Mifflin Company, Boston.

Money, Nicolas P. (2004) *Mr. Bloomfield's Orchard: The Mysterious World of Mushrooms, Molds, and Mycologists.* Oxford University Press, Oxford, UK.

Money, Nicolas P. (2017) *Mushrooms: A Natural and Cultural History.* Reaction Books, London.

Ostendorf-Rodríguez, Yasmine. (2023) *Let's Become Fungal! Mycelium Teachings and the Arts: Based on Conversations with Indigenous Wisdom Keepers, Artists, Curators, Feminists and Mycologists.* Valiz, The Netherlands.

Parkinson, Siôn. (2024) *Stinkhorn: How Nature's Most Foul-Smelling Mushroom Can Change the Way We Listen.* Sternberg Press, London.

Pegler, D. N., T. (1995) *Læssøe, and B. M. Spooner. British Puffballs, Earthstars and Stinkhorns: An Account of the British Gasteroid Fungi.* Royal Botanic Gardens, Kew, London.

Pollan, Michael. (2018) *How to Change Your Mind: What the New Science of Psychedelics Teaches Us About Consciousness, Dying, Addiction, Depression, and Transcendence.* Penguin Press, New York.

Rolfe, R. T., and F. W. Rolfe. (1925) *The Romance of the Fungus World.* Chapman and Hall, London.

Sheldrake, Merlin. (2020) *Entangled Life: How Fungi Make Our Worlds, Change Our Minds & Shape Our Futures.* Random House, New York.

Smith, Alexander. (1951) *Puffballs and Their Allies in Michigan.* University of Michigan Press, Ann Arbor, MI.

Sowerby, James. (1797-1809) *Coloured Figures of English Fungi or Mushrooms.* J. Davis, London.

Swanton, Ernest W. (1909) *Fungi and How to Know Them: An Introduction to Field Mycology.* Methuen, London.

Trinder, Kingston, et al. (2020) *John Cage: A Mycological Foray.* Atelier Éditions, Los Angeles.

Wasson, R. Gordon. (1973) *Soma: Divine Mushroom of Immortality.* Harcourt Brace Jovanovich, New York.

Woodward, Ben. (2012) *Slime Dynamics: Generation, Mutation, and the Creep of Life.* Zero Books, Winchester, UK.

ONLINE RESOURCES

American Phytopathological Society. apsnet.org/Pages/default.aspx.

Hunt Institute for Botanical Documentation. huntbotanical.org.

The Huntington Library, Art Museum, and Botanical Gardens: Library Collections: History of Science, Medicine, and Technology. huntington.org/history-science-medicine-and-technology.

Index Herbariorum. sweetgum.nybg.org/science/ih.

Mycological Society of America (MSA). msafungi.org.

NYBG: Fred Jay Seaver Records. nybg.org/library/finding_guide/archv/seaver_rg4f.html.

NYBG: LuEsther T. Mertz Library Plant & Research Guides. https://libguides.nybg.org/databases_resources.

USDA: Special Collections at the USDA National Agricultural Library. https://archivesspace.nal.usda.gov.

USDA: U.S. National Fungus Collections Topical Files. nal.usda.gov/collections/special-collections/us-national-fungus-collections-topical-files.

USDA: U.S. National Fungus Collections (BPI). ars.usda.gov/northeast-area/beltsville-md-barc/beltsville-agricultural-research-center/mycology-and-nematology-genetic-diversity-and-biology-laboratory/docs/us-national-fungus-collections-bpi.

USDA: Mycology and Nematology Genetic Diversity and Biology Laboratory. ars.usda.gov/northeast-area/beltsville-md-barc/beltsville-agricultural-research-center/mycology-and-nematology-genetic-diversity-and-biology-laboratory.

References

INTRODUCTION

1. Mycological Society of America, "About Us: The Mycological Society of America," msafungi.org/society
-information. Accessed November 2024.
2. Mycological Society of America, "About Us."
3. John A. Stevenson. Letter to Maurice B. Walters. 27 January, 1955.
4. Daniel J. Collins et al., "Contributions of Dr. George Washington Carver to Global Food Security: Historical Reflections of Dr. Carver's Fungal Plant Disease Survey in the Southeastern United States," *American Phytopathological Society (APS),* 1 February, 2014.
5. National Agricultural Library Special Collections, "U.S. National Fungus Collections Topical Files MS0454: Organizational History," 4.
6. Agricultural Research Service, US Department of Agriculture, "USDA Fungal Databases," fungi.ars. usda.gov. Accessed November 2024.
7. National Agricultural Library Special Collections, "U.S. National Fungus Collections Topical Files MS0454: Organizational History," 4.
8. National Agricultural Library Special Collections, "U.S. National Fungus Collections Topical Files MS0454: Summary Information," 3.
9. National Agricultural Library Special Collections, "U.S. National Fungus Collections Topical Files MS0454: Summary Information," 3.
10. Maurice B. Walters. Letter to John A. Stevenson. 4 February, 1955.
11. Maurice B. Walters. Letter to John A. Stevenson. 30 December, 1959.
12. Alexander H. Smith and Rolf Singer, "A Monograph on the Genus Cystoderma," *Papers of the Michigan Academy of Science, Arts, and Letters* 30 (1945): 129.
13. New York Botanical Garden, "Fred Jay Seaver Records (RG4): Biography of Fred Jay Seaver," nybg.org/library/finding_guide/archv/seaver_rg4f.html. Accessed November 2024.
14. Fred Jay Seaver, *The North American Cup-fungi (Operculates)* (1942), 345.
15. Seaver, *The North American Cup-fungi (Operculates),* 332.
16. Maurice B. Walters. Letter to John A. Stevenson, 6 July, 1941.
17. Maurice B. Walters, *"Pholiota Spectabilis,* A Hallucinogenic Fungus," *Mycologia* 57 (1965): 837.
18. Walters, *"Pholiota Spectabilis,"* 837.
19. Walters, *"Pholiota Spectabilis,"* 837.
20. Walters, *"Pholiota Spectabilis,"* 838.
21. Maurice B. Walters, "Wolffia Papulifera and Lemna Minima in Ohio," *Ohio Journal of Science* 50, no. 6 (November 1950): 266; Maurice B. Walters, "Mosses of a Northern Ohio Area," *Ohio Journal of Science* 52, no. 5 (September 1952): 291–95.
22. William C. Steere, "Notes on *Fissidens.* II. The Discovery of *Fissidens exilis* in North America," *The Bryologist* 53, no. 2 (June 1950): 131.
23. Steere, "Notes on *Fissidens.* II.," 136.
24. Joe Tait. Email to Maya Lydia Bushell, 5 December, 2024
25. Steere. "Notes on *Fissidens.* II.," 131.
26. Maurice B. Walters, US Draft Registration Card, 27 April, 1942.
27. Maurice B. Walters. Letter to Edith K. Cash. 22 October, 1947.
28. Maurice B. Walters, "Mosses of a Northern Ohio Area," *Ohio Journal of Science* 52, no. 5 (September 1952): 291.

THE GIANT PUFFBALL, THE AMATEUR, AND HIS LOVE OF DOGS

1. The figure of 117 to 125 billion, which is from the Population Reference Bureau, includes every human from prehistory to the present. See prb.org/focus
-areas/world-us-population-trends.

SPORE TIME

1. The first printed illustration of a mushroom appeared in the *Ortus sanitatis* (also known as the *Hortus sanitatis,* or *Garden of Health*), book 1, plate CCCIII, published in 1491 by an anonymous author.

SIÔN
PARKINSON

Siôn Parkinson is an artist, musician, and writer investigating our sensory relationship with the more-than-human world. His first book, *Stinkhorn: How Nature's Most Foul-Smelling Mushroom Can Change the Way We Listen* (2024), invites readers to discover the musical potential of unpleasant odors by "listening through the nose." Siôn is a research fellow at the Royal Botanic Garden Edinburgh, where he is exploring the olfactory heritage of fungi—mushroom smells that are meaningful to individuals or communities due to their association with significant places, objects, or traditions. He is based in Dundee, Scotland.

MAYA
LYDIA BUSHELL

Maya Lydia Bushell is a freelance writer, musician, and editor based in Los Angeles. She is a graduate of Trinity College, Dublin, with a TSM Hons degree in English Literature and History of Art & Architecture (2020). In 2021, she completed a residency with Can Serrat (El Bruc, Spain) for which she produced a dual field recording and text-based project, *Impresiones de El Bruc*, which was subsequently released by Jungle Gym Records. Her discography now includes four projects with JGR, and her music has been licensed for use in the television show *Devil in Ohio* (Netflix, 2022). She has been an associate at Marta, a gallery for art and design in Los Angeles, since 2022.

PERMISSIONS

ALL IMAGES

US National Fungus Collections Topical Files, Mycological Illustrations and Photographs, Maurice B. Walters Collection, 1935–1952. Courtesy of Special Collections, USDA National Agricultural Library.

About the Collection

The US National Fungus Collections is a mycological institution that includes the Western Hemisphere's largest fungal herbarium (Herbarium BPI), the John A. Stevenson Mycological Library, the USDA Fungus-Host Databases, and the only actively curated nomenclature database focused on plant pathogenic fungi. It is a base for foundational research and service in national and international mycology and plant pathology.

The Maurice B. Walters Photographs are held in the US National Fungus Collections Topical Files, which contain historical records assembled largely by John A. Stevenson during his 33 years of service as director of the US National Fungus Collections. These materials reflect the history of American mycology and plant pathology and the relationship to the development of the US National Fungus Collections in the 19th and 20th centuries. These files include correspondence; biographical information; unpublished manuscripts; information on scientific meetings, other herbaria, associations and societies; mycological and phytopathological data; photographs; and field and laboratory records.

They are kept in the Special Collections of the National Agricultural Library (NAL), one of five national libraries of the United States. As part of the USDA and the Agricultural Research Service (ARS), the library houses one of the world's largest collections on agriculture and its related sciences. NAL's Special Collections unit preserves and provides access to materials significant to the history of agriculture and the USDA, including rare books and manuscripts, document collections, nursery and seed trade catalogs, photographs, and posters.

Search the complete collection:
search.nal.usda.gov/discovery/
collectionDiscovery?vid=01NAL_
INST:MAIN&collectionId=81490462630007426

Colophon

UNION SQUARE & CO. and the distinctive Union Square & Co. logo are trademarks of Hachette Book Group, Inc.

© 2025 UNION SQUARE & CO.

All Artworks by Maurice B. Walters
Introduction by Maya Lydia Bushell
Additional Essays by Siôn Parkinson

ISBN 978-1-4549-6355-4

Union Square & Co. books may be purchased in bulk for business, educational, or promotional use. For more information, please contact your local bookseller or the Hachette Book Group's Special Markets department at:
special.markets@hbgusa.com.

Printed in Spain

10 9 8 7 6 5 4 3 2 1

unionsquareandco.com

atelier éditions

A book by Atelier Éditions
atelier-editions.com

Editor: Pascale Georgiev
Contributing Editor: Maya Lydia Bushell
Editorial Assistant: Agnes Perotto-Wills
Designer: Emma Singleton
Art Director: Renée Bollier
Project Editor: Caitlin Leffel

The information in this book is meant to illustrate one mycologist's collection of field studies and should not be used as a field guide or identification tool. We recommend readers apply caution when foraging for wild mushrooms, especially if untrained.

ACKNOWLEDGMENTS
Atelier Éditions wish to extend their immense gratitude to the National Agricultural Library of the US Department of Agriculture, especially to Sara Lee, Clayton Ruminski, Timothy Schoepke, and Matthew Pearson and the team at the Special Collections Department, who helped bring this project together. The team also wish to thank Maya Lydia Bushell for bringing us her vision of making a mushroom follow-up to *An Illustrated Catalog of American Fruits & Nuts*.

We thank Dr. Lisa A. Castlebury and Shannon Dominick at the Mycology and Nematology Genetic Diversity and Biology Laboratory of the USDA; Daniel Sone, photographer and digitizer at the USDA; the USDA Metadata team, led by Clayton Ruminski and including Michelle McDaniel, Amy Morgan, Jesse Padron, Kelvin Posey, Amanda Ray, Daniel Weddington, and Raymond Williams, and Timothy Schoepke for his reference work; Stephen Sinon at the LuEsther T. Mertz Library of the NYBG; Dr. Meredith Blackwell and Dr. Donald H. Pfister at Louisiana State and Harvard Universities, respectively, for their aid and insight into the MSA archives; and Joe Tait at the Cleveland Museum of Natural History.